Visualization in Supercomputing

Raul H. Mendez
Editor

Visualization in Supercomputing

With 166 Illustrations, 25 in Color

Springer-Verlag
New York Berlin Heidelberg
London Paris Tokyo Hong Kong

Raul H. Mendez
Institute for Supercomputing Research
15F Inui Building Kachidoki
1-31-1 Kachidoki, Chuo-ku
Tokyo 104
Japan

About the cover:
The object is the quaternion Julia set for the function $z 2 + 0.2809 + 0.53i$. The upper front left octave is missing, revealing the basins of attraction of the four-cycle defining its interior. These interior points are such that they remain bounded under repeated applications of the function. In fact, they are periodic, moving from red to green to yellow to blue to red again. The image was ray-traced on an AT&T Pixel Machine 964dX by John C. Hart at the Electronic Visualization Laboratory at the University of Illinois at Chicago. Copyright 1990 John C. Hart.

Library of Congress Cataloging-in-Publication Data
Visualization in supercomputing/editor, Raul H. Mendez.
    p. cm.
   ISBN-13:978-1-4612-7971-6
   1. Supercomputers. I. Mendez, R. (Raul)
  QA76.5.V55   1990
  004.1'1—dc20   89-48589

Printed on acid-free paper.

Camera-ready copy prepared by authors.

9 8 7 6 5 4 3 2 1
ISBN-13:978-1-4612-7971-6   e-ISBN-13:978-1-4612-3338-1
DOI:10.1007/978-1-4612-3338-1

Preface

In August 1988, one of the finest groups of supercomputing and visualization researchers from the U.S. and Japan met in Tokyo for a four-day conference. This conference was unique in that it was the first gathering of the countries' leading scientists in computer visualization and high-speed computing. It was also one of the rare opportunities for visualization scientists and hardware designers to discuss and exchange views in this exciting field.

The idea of visualization is not new; because massive amounts of numeric data are far more comprehensible when converted into graphical form, visualization is becoming an intimate part of many areas of research. Techniques for visualization, however, are still being developed, and visualization research is just beginning to be recognized as a cornerstone of future computational science. As scientists bring increasingly complex problems to computational science, visualization will become an even more essential tool for extracting science out of numbers.

The goal of this conference was to provide a focal point for users and hardware/software designers, scientists from both ends of the visualization spectrum, to share their progress and discuss their needs. The conference site itself, one of the world's centers of computer technology, provided an auspicious environment for a successful conference. The present volume is a collection of papers selected from those presented at the conference. It is divided into three parts, representing the major areas in visualization computing: visualization applications, hardware/performance, and visualization theory.

In Part 1, Visualization Applications, Kozo Fujii's paper reviews visualization methods used in computational fluid dynamics research. In particular, he discusses the hardware requirements for doing effective viusalization of fluid dynamics from the researcher's point of view. The time-to-solution, an often overlooked aspect of visualization, is addressed in Kumazawa's paper. In this article, various system configurations are examined to study the turnaround time from computer simulation to grahpical display. Having the fastest computer does not necessarily mean the shortest time-to-solution.

Part 2, covering visualization hardware and performance, includes a paper by Robert Fulton and Kuo-Ning Chiang on the use of parallel/vector computers with finite-element systems. The paper by Kok-Meng Lue and Kazuto Miyai addresses the basic computational performance of two graphics supercomputers, the Ardent Titan and Stellar GS1000.

In Part 3, Alvy Ray Smith's paper covers important topics such as volume and 3D surface imaging, geometry-based graphics workstations vs. image computers. The paper also discusses the applicability of the volume imaging concept in various fields.

Lastly, I have collected in the appendix viewgraphs from several presentations at the conference which should complement the materials presented.

Raul H. Mendez
Tokyo, March 1989

Contents

Part 1

Visualization Applications

Supercomputers and Workstations
in Fluid Dynamic Research

Kozo Fujii

Associate Professor, High Speed Aero. Division
The Institute of Space and Astronautical Science,
Yoshinodai 3-1-1, Sagamihara
Kanagawa, 229 Japan

Abstract

Computational Fluid Dynamics (CFD) is beginning to play an important role in the aeronautical industry all over the world. People now realize that CFD can be a new and effective design tool. One of the important factors that has been accelerating the development of CFD was rapid progress of supercomputers. With the increase of the data obtained by the CFD research, importance of visualization of the computed results began to be recognized. Graphic workstations that were also rapidly progressing then became another key factor for CFD research. In the present paper, how supercomputers and graphic workstations are used in the CFD research is discussed. The requirements for these machines from the view point of CFD research are also addressed.

Key issue for the graphic workstation is the software development. Such softwares should be interactively used without requiring researcher's effort and can display the plots they want quickly. Displaying beautiful and real-image pictures may be useful but does not have the first priority. The network between the supercomputers and graphic workstations is also important.

1. Introduction

The research area that studies fluid dynamics using computers is rapidly growing. This research area is called 'Computational Fluid Dynamics (CFD)' and is becoming a strong design tool for aircraft industries as well as becoming a good tool for understanding of fluid physics. One of the strong factor that is accelerating CFD

3

research is the appearance and rapid progress of supercomputers. The computed result usually becomes more accurate and reliable by increasing the number of grid points. Thus, supercomputers with high-speed operations and large memory are suitable for flow simulations, and now are inevitable. To help people to understand what happens in the flow field from the obtained data, we have to visualize the flow and the use of graphic workstations has an important role for that purpose. Compared to the circumstances several years ago, the level of the graphic software is improved, and visualizing complex flow field in realistic image now is not a difficult task.

Thus, both supercomputers and graphic workstations are key hardwares for producing advance CFD results. At the same time, we have to develop good softwares. The algorithms for the flow simulations are still being improved year by year. Developing the software to visualize the computed flow field on the workstations is also important. 'Visualization' itself is important in many engineering fields such as commercial films, medical sciences, and structural design. There are many common requirements of these visualization and they are mostly applicable to the CFD visualization. However, there exist several points we should remember when considering the CFD visualization.

In the present paper, how supercomputers and graphic workstations are currently used in the CFD is first described. Then, what kind of capability is required on supercomputers and graphic workstations is discussed. Recent trend on the usage is demonstrated and what will be needed in the future is finally discussed.

2. Computational Fluid Dynamics

The CFD study consists of many sub-research areas. They are: algorithm development for flow simulation, application of the developed computer codes for the simulation, evaluation of the applicability of the developed codes by the comparison with the experiment, and so on. Basically, there are two useful applications of CFD.

First one that has been actually accelerating the CFD is the engineering use. In the process of designing new aircraft, for instance, people have been conducting a lot of wind tunnel testing, where many small-scaled model are tested to find the best configuration. The wind tunnel testing requires large man and electric power and thus takes a long time. Computer simulation (CFD), on the other hand, gives us large amount of data within reasonable time. The CFD has been intensively used for aircraft design especially in the initial stage and is now inevitable. The use of the CFD is also becoming popular in another engineering fields. People begins to use the CFD for the design of turbines, motor vehicles, and many other engineering products.

The second application is basic physical study of fluid dynamic phenomena. Assuming that the CFD is reliable, we can discuss fluid physics from the computed results. In the computation, we can plot the resultant data in any form. For instance, we can plot density, pressure, particle path etc. and even vorticity or entropy distributions can be plotted. Besides, it is easy to stop the computation and see one instantaneous solution. These are not necessarily easy in the experiment. In many cases the flow is sensitive and tends to become totally different by the interference of the probe inserted to measure the flow field. We can look at more detail of the flow structure using the CFD.

Note that there is one big difference between the first application and the second application described above. In the first application using the CFD as a design tool, there is a restriction of the time span. Many solutions should be supplied within reasonable time and expense. The solutions should be validated but the accuracy can be the second requirement. In the second application to study fluid physics, obtaining accurate solution is most important and computer time could be large.

What made the CFD to be so popular? As shown in the examples above, it can be summarized as 1) many solutions (flow fields) can be obtained with relatively little time and expense, 2) changing the geometry is easy, 3) a lot of information can be deduced, 4) the computer code is portable and can be run on any computers. These facts enable us to try risky development such as spaceplane. However, these benefits of the CFD are realized only when both supercomputers and graphic workstations are available.

3. Flow Simulations on Supercomputers

Let's discuss how we conduct flow simulations on computers to explain why supercomputers are so important for the CFD. When tracing the trajectory of the moving body, we solve the Newton's second law of motion. Similarly, when we want to know fluid motions, we solve the equations to rule (at least believed to rule) fluid motions. Unfortunately, these equations are system of nonlinear partial differential equations, and various approximations have to be introduced to make the equations simpler. The equations having been used for the real design of aircraft are the equations with many approximations because certain amount of solutions have to be obtained within required time span. However, with the rapid progress of supercomputers research focus now is to solve the equations with little approximation. These equations are called 'Navier-Stokes equations' that can describe flow fields accurately. Large computing time and computer memory are required since the equations have little approximation and thus are complicated.

Discretization method such as finite difference or finite volume approaches is mainly used to solve these equations. The flow field to be solved is divided into many cells (we call these computational grids). We discretize the equations to be solved and consider the discretized equations for each cell. Then the result is obtained as flow variables at each cell. In other words, what we obtain is the flow variables such as density, pressure velocities, etc. at each cell. As expected, the result becomes more accurate by simply increasing the number of cells. Since we consider the equations represented by each cell, the required computer memory is then increased and so is the computer time. Appearance of supercomputers with large memory and fast processors really helped this problem. For instance, two-dimensional Navier-Stokes solutions for a transonic airfoil could be obtained within a couple of minutes on the current supercomputers. Similar computations would require more than one hour on typical conventional computers. The effect is pronounced for the three-dimensional computations. Figure 1 shows the pressure contour plots over the transonic transport aircraft configuration. The computation[1] was carried out in 1985 on one of the Japanese supercomputers, and it required 4 to 6 hours of computer time and 256 Mbytes memory. Our experience[2] on the code performance indicated that the computation would have taken more than 300 hours even though we did not try it because scalar computers with such large memory was not available. From our recent benchmark test, we could say that similar computation would take less than one hour on the coming supercomputers. Although the required computer time is still not small enough for the design purpose, and further improvement of the supercomputers as well as the algorithms improvement is required, it is very true that supercomputers are key factor for the computational fluid dynamic research.

4. Computational Fluid Dynamics and Graphic Workstation

Now, let's turn to graphic workstations. As stated in the previous section, the computed results are obtained in the form of physical variables for each cell. Suppose we only had the printout of the numbers showing the physical variables and the associated coordinates, we could never imagine the flow field. It is especially true for the three- dimensional flow fields. Therefore we have to show the result in the fashion that people can understand it easily. Even with good supercomputers we could never understand flow physics suppose we did not have tools to visualize the results. This is also true for the experimental fluid dynamics, and has been called 'flow visualization'. What we have to do for the computational fluid dynamics is the 'computational flow visualization'. Actually, Figure 1 is one of the visualized output from the computed result. It may not be clear, but fluid dynamicist can tell much from this figure.

What is required for the 'computational flow visualization'? Figure 2 shows the total pressure contour plots for the high-angle of attack flow over a strake-delta wing which represents the wing configuration for fighters, supersonic transport aircraft and spaceplanes. Strong vortex region is presented as a circle-shape dense contour lines in this type of plot. Such region (looks like eye-shape in the figure) is recognized from the front above the wing, and there observed is a large hole in this region near the wing trailing edge. Fluid dynamicists know that vortical flow exist along the circle-shape region, and thus interpret that something, that changes this vortical flow is happening near the trailing edge. To find out what happens there, they want to see some other plots, such as density, pressure, vorticity. This figure shows contours in seven streamwise planes, but the contours in the other planes may also be required. Furthermore, other functions such as changing the view point, zooming up the plot, etc. may also be helpful. Figure 2 shows one image of the postprocessor we have developed on the mainframe computer for the quick-look purpose. In this plotting package, viewing positions, contour levels and the sections to display can be changed interactively.

These functions are important for understanding flow fields. The plot need not be beautiful, but should be quickly displayed on the screen as soon as the solution is obtained. At the same time, many functions that are required by fluid dynamicists should be displayed. Many visualizations require realistic and beautiful picture images, but it in not too important for the flow visualization.

Which machines are suitable for computational flow visualization? To survey the field of computers, workstations are in trend. There are plenty of workstations and some have so high CPU performance as large-scale computers of one generation ago although they were originally developed mainly for distributing tasks and reducing the load of host computer. Among them, Graphic Workstations (GWS) and super GWS are focused because they have high performance of graphics as well as that of floating point calculations. Such GWS is suitable because of their high performance of graphics and refined man-machine interface (MMI). In fact, Such visualization system has already been developed on the IRIS workstations at NASA Ames Research Center[3,4].

Since recent CFD research is focused on the three-dimensional computations, 3-D rendering capability is important. Figure 3 shows the color image of the similar plot that the author obtained at NASA Ames Research Center two years ago[5]. If the color table is prepared (hopefully displayed on the screen too), we can tell the level of the contours. The color image also distinguishes the wing configuration with the contour plots. Thus, even though displaying 'realistic and beautiful' image on the screen does not have the first priority, plotting the data with color maps helps understanding of the computed result. Another example from the same computational data is shown in Fig. 4. This figure shows the so called particle paths, where we find the flow directions (streamline patterns). Tightly-coiled vortex (yellow and orange straight lines from the top) emanating from the leading edge suddenly blows up (blue region), and we know more clearly that something is happening in the region over the aft portion of the wing. Since the color denotes the local density of the flow, we also know that the density also suddenly changes there. This is the phenomenon called 'vortex breakdown' and the egg-shaped jammed region is created.

We have made the animation based on this result at NASA Ames Research Center. The animation told us that there was a clear ordered motion inside this breakdown region. On the other hand, we only tell that complicated fluid motion exists inside the breakdown region from the still picture. That is an example to explain why animation is important for the CFD research. However, again it is worthless if creating the animation requires long time, say one week. We need to see the animation (on the display screen) as soon as the result is obtained.

Computer flow visualization requires good graphic workstations to help our understanding of the fluid motion, but it is just a tool for the fluid dynamic research. Therefore, most important point is that the visualization process should not require researcher's intensive effort. Good softwares in addition to good hardwares are important.

4. Further Requirements on Flow Visualizaion

The computer programs that satisfy some of the requirement mentioned above have been developed at NASA Ames Research Center. They are called 'PLOT3D', 'GAS' and 'RIP' and each of them has its own unique feature[4,5]. Figure 4 was obtained using 'RIP' where the particle trajectories were computed on the CRAY-2 supercomputer and the result was displayed on the IRIS workstation. The graphic process was carried out interactively on the IRIS display. The network connecting the supercomputer and the workstation and its speed were critically important for the work.

Currently, flow computation is usually conducted on supercomputers, and the result is displayed on the workstation screen after sending the result from the supercomputer to the workstation. Since the data to be transferred is becoming larger and larger, the speed of the network is important here again. The CFD now focuses on the steady-state flow computation (steady-state means that the solution does not change in time) because of the supercomputer speed and memory. New trend is to compute the unsteady flow. The computed result then includes the time sequence of the solution and becomes enormous. Suppose we want to monitor the solution evolving in time on the supercomputer using the workstation's display, the network has to be much more faster. Another option is to compute the flow on the workstation as well as displaying the result. When considering the rapid progress of superworkstations, it may be realized in the near future especially for simple problems. At least, we have to remember there is such option and try to find the problems associated with that option.

As graphic workstations become popular, we are frequently asked which workstations are good to buy. The answer is of course difficult, but we have to remember that the answer depends on how workstations are going to be used. They have to buy the workstations having good computer capability if they want to compute the flow on the workstation. The choice may be the workstations having good graphic capability if they compute the flow on supercomputers and displaying the result on the workstations. It also depends if they are interested in making nice and beautiful pictures for the presentation (this is not emphasized in this paper, but it is a useful tool to get budget etc.), or if they are interested in fluid dynamic research. The software to be developed also becomes different based on this choice.

One of the super Graphic Workstation has been introduced at our laboratory lately and the postprocessor (flow visualization programs) for Navier-Stokes and other solvers having the features discussed in this paper are in the process of being built up on that machine[6]. Figure 5 shows the overview of the system that we are developing.

5. Summary

Computational Fluid Dynamics (CFD) is beginning to play an important role in the aeronautical industry all over the world. As the problems dealt with the CFD research become practical and large scale, use of supercomputers for the computation and the workstations for the visualization become inevitable. In the present paper, supercomputers and graphic workstations in the CFD research were discussed and the requirement by the CFD research was clarified. Key issue for the graphic workstation is the software development. Such softwares should be interactively used without requiring researcher's effort and can display the plots they want quickly. Displaying beautiful and real-image pictures may be useful but does not have the first priority. The network between the supercomputers and graphic workstations is also important.

References

1. Fujii, K. and Obayashi, S., "Navier-Stokes Simulations of Transonic Flows over a Wing-Fuselage Combination," AIAA Journal, Vol. 25, No. 12, December, pp.1587-1596, 1987.
2. Fujii, K., "Recent Application of the LU-ADI Scheme to the Viscous Compressible Flow Simulations," *Lecture Notes in Engineering*, No. 24, Supercomputers and Fluid Dynamics, Springer-Verlag, January, 1986.
3. Lasinski, T., et. al., "Flow Visualization of CFD Using Graphics Workstations," AIAA Paper 87-1180, 1987.
4. Watson, V., et. al., "Use of Computer Graphics for Visualization of Flow Fields," presented at AIAA Aerospace Engineering Conference and Show, February 17-19, 1987.
5. Fujii, K. and Schiff, L. B., "Numerical Simulations of Vortical Flows over a Strake-delta Wing," AIAA Paper 87-1987, 1987, to appear as AIAA Journal article in 1989.
6. Tamura Y. and Fujii, K., "Use of Graphic Workstations for Computational Fluid Dynamics," submitted for the presentation at the International Symposium on Computational Fluid Dynamics-Nagoya, to be held at Nagoya, Aug., 1989.

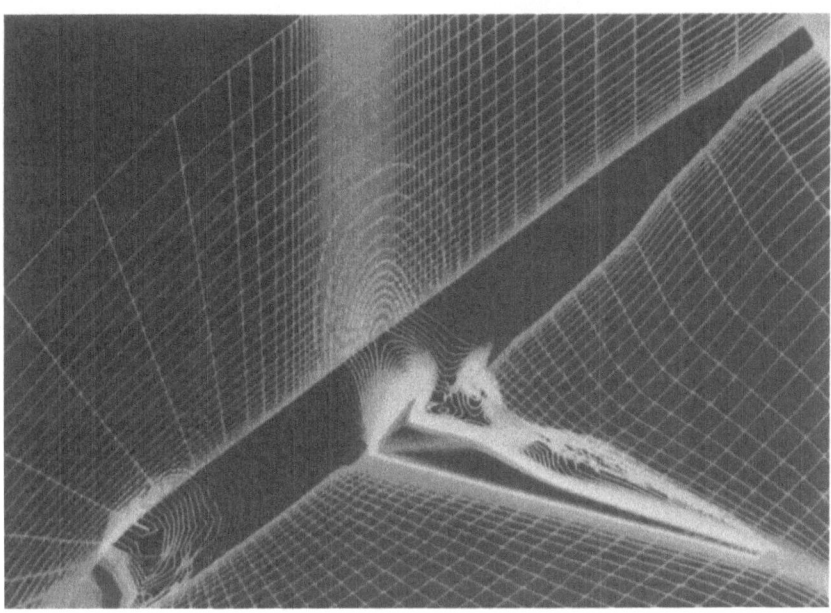

FIGURE 1. Pressure contour plots on the transport aircraft. (Color art for this figure may be seen in the color insert.)

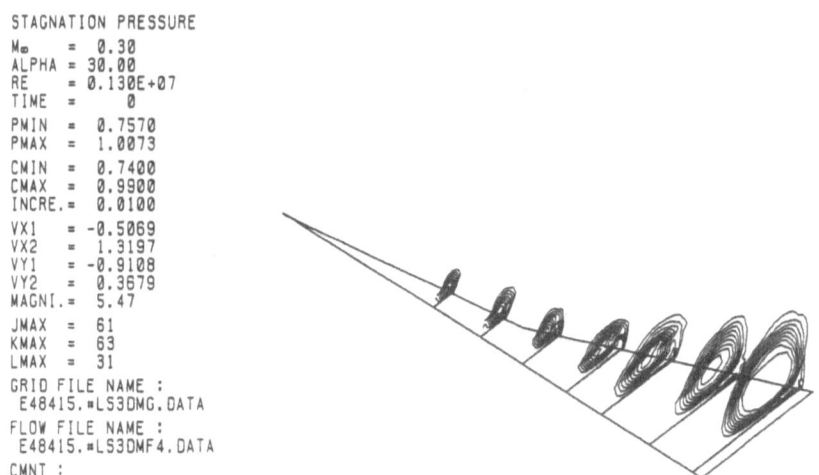

```
STAGNATION PRESSURE
M∞    =   0.30
ALPHA = 30.00
RE    = 0.130E+07
TIME  =      0
PMIN  =   0.7570
PMAX  =   1.0073
CMIN  =   0.7400
CMAX  =   0.9900
INCRE.=   0.0100
VX1   =  -0.5069
VX2   =   1.3197
VY1   =  -0.9108
VY2   =   0.3679
MAGNI.=   5.47
JMAX  =  61
KMAX  =  63
LMAX  =  31
GRID FILE NAME :
 E48415.*LS3DMG.DATA
FLOW FILE NAME :
 E48415.*LS3DMF4.DATA
CMNT :
```

FIGURE 2. One image of the flow visualization program: Vortex breakdown over a · strake-delta wing.

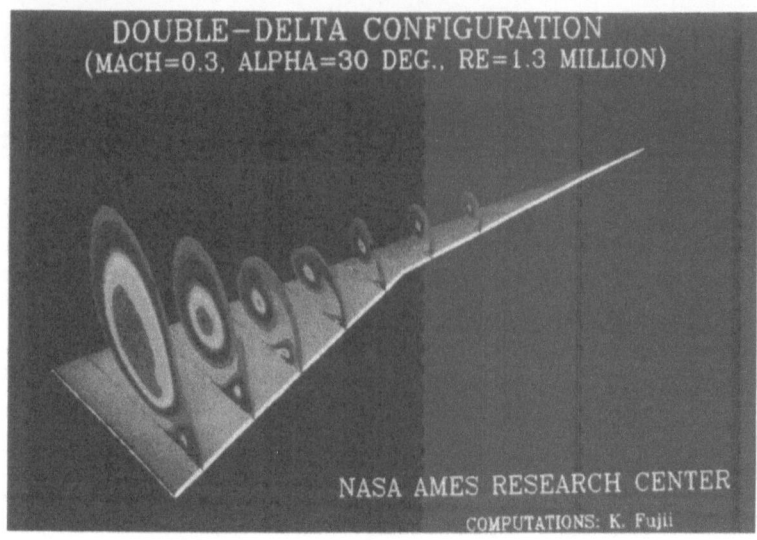

FIGURE 3. Color image of contour plots: Vortex breakdown over a strake-delta wing. (Color art for this figure may be seen in the color insert.)

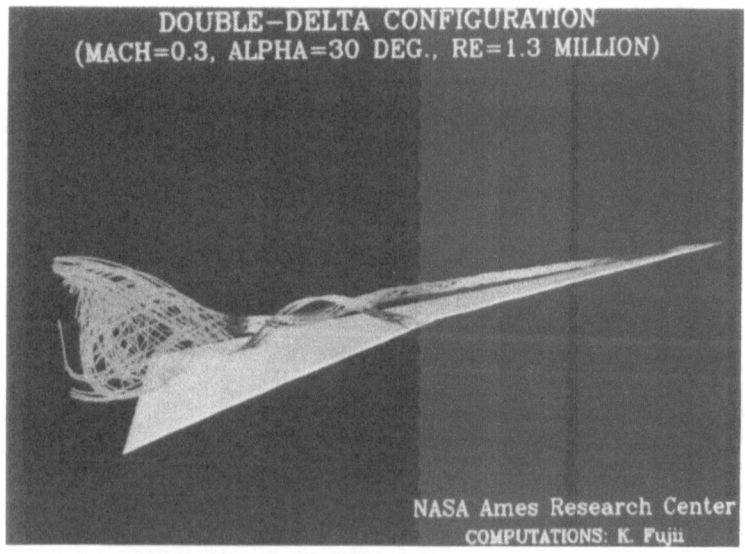

FIGURE 4. Particle paths showing vortex breakdown. (Color art for this figure may be seen in the color insert.)

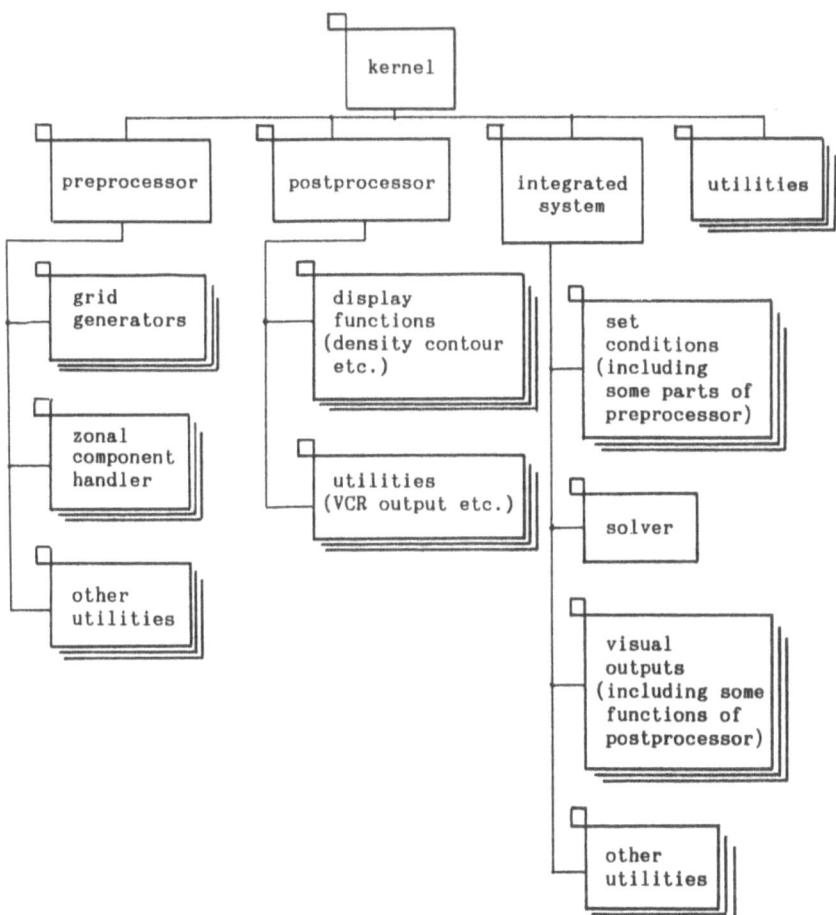

FIGURE 5. Overview of the CFD supporting system under development.

Numerical Simulation of A 3-Dimensional Backward-Facing Step Flow

Hiroshi TAKEDA and Erika MISAKI

Institute for Supercomputing Research, Recruit Co., Ltd,
Recruit Kachidoki Bldg., 2-11, Kachidoki, Chuo-ku, Tokyo 104, Japan

Abstract

Numerical simulation of a backward-facing step flow for Re = 800, 1200, and 1600 was performed to investigate the three-dimensionality of flows in the transition Reynolds number range. Through flow visualization we conclude that the flow is almost steady and 2 dimensional at a Reynolds number of 800. However, for Re = 1200 and 1600, vortices with 3-dimensional structures are generated periodically and advected downstream along the bottom and top walls.

Introduction

The separation and reattachment of a shear flow layer is an important process in many flow related engineering problems, for example, diffusers, airfoils, buildings, and internal combustors. To understand these flows, we need to understand and be able to predict the behavior of separating and reattaching shear layers.

Among such flows, a backward-facing step flow is the simplest and has most frequently been investigated both experimentally and numerically. Previous numerical simulation of the backward-facing step flow has been classified into two categories; simulation for laminar flow and simulation for turbulent flow. In the turbulent flow, the statistical flow fields are almost 2-dimensional and their simulation has been performed mainly for the study of turbulence models. On the other hand, the laminar-flow

14

simulation has been performed for computer-code validation and to study the physics of separation and reattachment of shear flows. Unfortunately, only 2-dimensional flows with the Reynolds number up to 800 have been previously studied (Kim and Moin, 1985; Sohn, 1988). According to experimental results (Armaly et al., 1983), the backward-facing step flow becomes 3 dimensional but remains laminar for Reynolds number between several hundreds and several thousands. This Reynolds number range is called the 'transition range' and is physically important. In this study, we simulate the flow in the transition Reynolds number range using a 3-dimensional code to investigate the flow behavior.

With the advent of computer graphics, supercomputer visualization plays an equally important role as the simulation itself in understanding the physical phenomena. In 2-dimensional simulations, streamline contours is usually the most useful for visualization of the data. However, streamlines cannot be defined in general 3-dimensional flows, and hence, other effective visualization techniques must be introduced. We therefore try to visualize the data by using the vorticity contours, pressure contours, time line of pressure gradient, and streak lines.

Outline of Computation

Figure 1 shows the geometry of the computational region in Cartesian coordinates. The computational region has been defined in such a way that it coincides with the experimental region in Armaly et al. (1983). The flow has been confined by two side-walls with width of 36 in the spanwise direction in the experiments. In the computation, we assume the flow being symmetrical about $z = 0$ and apply the nonslip wall condition at $z = 18$. Although another computation was done with another wall at $z = -18$, we have obtained the same result as in the symmetric case. In addition, the experiments show that the flow is symmetric in the spanwise direction with the Reynolds number up to a few thousands. Consequently, this symmetric assumption is sufficient. The outflow boundary condition is applied at $x = 40$.

The computation is performed using a recently developed finite difference scheme by one of us (Takeda, 1989). The scheme is of second-order accuracy in both space and time. The number of grid points is 201 x 41 x 11 (i.e., $\Delta x = 40/200$, $\Delta y = 2/40$, $\Delta z = 18/10$) and $\Delta t = 4\Delta y = 0.2$ was used as the time interval.

Numerical Results

We performed computations for Re = 800, 1200, and 1600, where the Reynolds number is defined with the step height and the two thirds of the maximum (i.e., mean) inflow velocity.

Figure 2 shows contours of the z-component of vorticity in the x-y plane (0 ≤ x ≤ 40, 0 ≤ y ≤ 2) at z = 0 (lower figure) and 9 (upper figure) at t = 1000, where contours with negative values have been plotted using dotted line and a contour interval of 0.5. Figure 3 shows contours of the x-component of vorticity in the z-y plane (0 ≤ z ≤ 18, 0 ≤ y ≤ 2) at x = 10, 20, and 30 at the same time step. The contour interval is 0.01 for figures of Re = 800 and 0.2 for figures of Re = 1200 and 1600. Figure 4 shows pressure contours in the x-z plane at y = 0 (lower figure) and y = 2 (upper figure) at t = 1000. It is seen in these figures that the flow is almost 2 dimensional in the case of Re = 800 and tends to become 3 dimensional as the Reynolds number increases. When the Reynolds number is 800, we notice two vortices in Fig. 2: a clockwise vortex behind the step and a counter-clockwise vortex along the top wall; these vortices have been reported in the experiments. On the other hand, multiple vortices can be found along the bottom and the top walls at Reynolds number 1200 and 1600. In the experiments, the second vortex along the bottom wall has been reported, but existence of multiple vortices was first detected in the present study. It is found in Figs. 3 and 4 that these vortices are 3 dimensional.

Figure 5 shows time variation of the pressure gradient in the x-direction (0 ≤ x ≤ 40) for t = 1000 to 1020. It is seen that the several vortices along the bottom and top walls detected in Figs. 2 - 4 are generated around the reattachment point of the primary vortex behind the step and are advected downstream with almost constant velocity. When Re = 1600, the generation point of the vortices becomes close to the step and the reattachment point also becomes close to the step compared with the case of Re = 1200. This is supposed to be due to the fact that the primary vortex behind the step has become more unstable for Re = 1600 and easily releasing small vortices in the downstream direction, i.e., small vortices are released nearer the step. On the other hand, the two vortices in the case of Re = 800 are found to be almost steady.

Figure 6 shows the x coordinate of the reattachment point of this stationary vortex behind the step as a function of z in the case of Re = 800. Although three-dimensionality has hardly appeared in Figs. 2 - 5 for Re = 800, it has appeared perceptibly in the reattachment point. We have also performed the 2-dimensional computation for Re = 800. According to that, the reattachment point is 12 and agrees with the numerical result of Kim and Moin (1985). The length of the primary vortex observed in the experiment is 14 and coincides approximately with the result in Fig. 6.

Figure 7 shows the side and top views of streak lines at t = 1000. It can be seen in the pictures of Re = 800 that the existence of the vortex behind the step and the vortex along the top wall. The three-dimensionality of the flows for Re = 1200 and 1600 can be observed as well.   Especially, it can be seen in the case of Re = 1200 and 1600 that the streak lines in the vortex behind the step roll greatly also in the spanwise direction.

The above computation was performed on a SX-2A supercomputer and the CPU time for each computation is between 4 - 8 hours depending on the the Reynolds number.   Figure 7 was drawn on a TITAN graphic supercomputer.

Conclusions

Numerical simulation of the backward-facing step flow for Re = 800, 1200, and 1600 was performed to investigate the three-dimensionality of the flow at the transition Reynolds number range using a 3-dimensional finite difference code.   Through the use of visualization techniques the following conclusions were obtained. For Re = 800, two vortices exist behind the step and along the top wall; they are almost steady and 2 dimensional. For Re = 1200, vortices with 3-dimensional structures are generated periodically around the reattachment point of the primary vortex behind the step and are advected downstream along the bottom and the top walls. For Re = 1600, such vortices become more unsteady and 3 dimensional. As a result, they are generated at points closer to the step and the reattachment point of the primary vortex becomes closer to the step. The unsteady and 3-dimensional vortices advected downstream were detected for the first time in the present study.

In our data visualizing the vorticity contours of the spanwise and streamwise components, the pressure contours, the time line of pressure gradient in the streamwise direction, and the streak line were used. Through these visualization methods, the flow patterns could be understood fairly well.   However, they are not sufficient to catch details of the 3-dimensional flow patterns, such as 3-dimensional structures of the unsteady vortices advected downstream. Other visualization techniques will be necessary for understanding them. For example, 3-dimensional plots of the velocity vector and contour surfaces of the pressure and the vorticity in 3 dimension may be helpful in understanding these complex structures, although it will be not easy to plot them. This 3-dimensional visualization is a problem to be considered hereafter.

18

References

B. F. Armaly, F. Durst, J. C. F. Pereira and B. Schönung: J. Fluid Mech., **127** (1983), 473-496.
J. Kim and P. Moin: J. Comput. Phys., **59** (1985), 308-323.
J. L. Sohn: Intern. J. Numerical Methods in Fluids, 8 (1988), 1469-1490.
H. Takeda: ISR Tech. Rep., TR89-6 (1989).

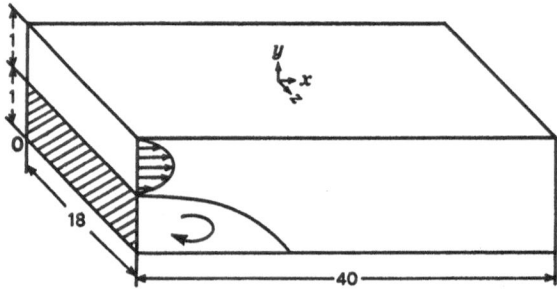

Fig. 1. Geometry of the computational region in Cartesian coordinates.

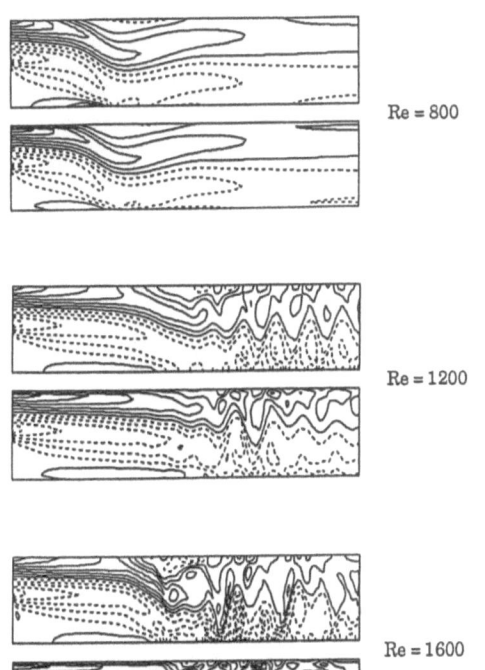

Re = 800

Re = 1200

Re = 1600

Fig. 2. Contours of the z-component of vorticity in the x-y plane ($0 \leq x \leq 40$, $0 \leq y \leq 2$) at z = 0 (lower figure) and 9 (upper figure) at t = 1000, where contours with negative values have been plotted using dotted lines and a contour interval of 0.5.

Re = 800   Re = 1200   Re = 1600

x = 30

x = 20

x = 10

Fig. 3. Contours of the x-component of vorticity in the z-y plane ($0 \leq z \leq 18$, $0 \leq y \leq 2$) at x = 10, 20, and 30 at t = 1000. The contour interval is 0.01 for figures of Re = 800 and 0.2 for figures of Re = 1200 and 1600.

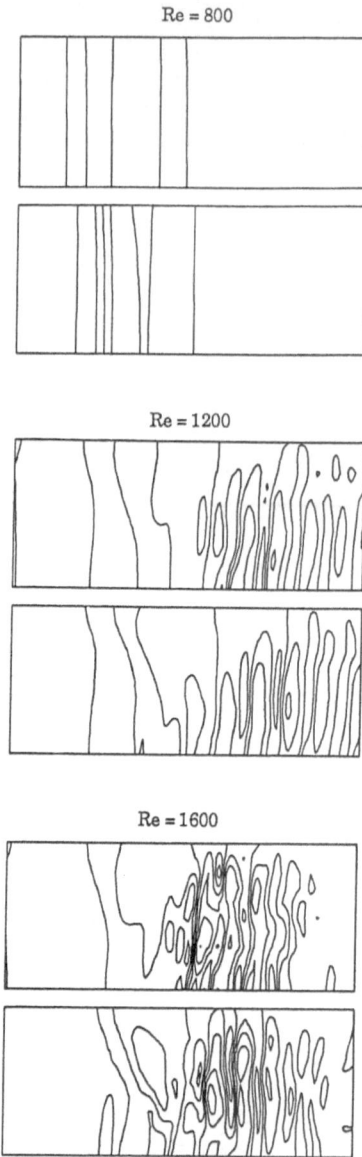

Fig. 4. Pressure contours in the x-z plane ($0 \leq x \leq 40$, $0 \leq z \leq 18$) at $y = 0$ (lower figure) and $y = 2$ (upper figure) at $t = 1000$.

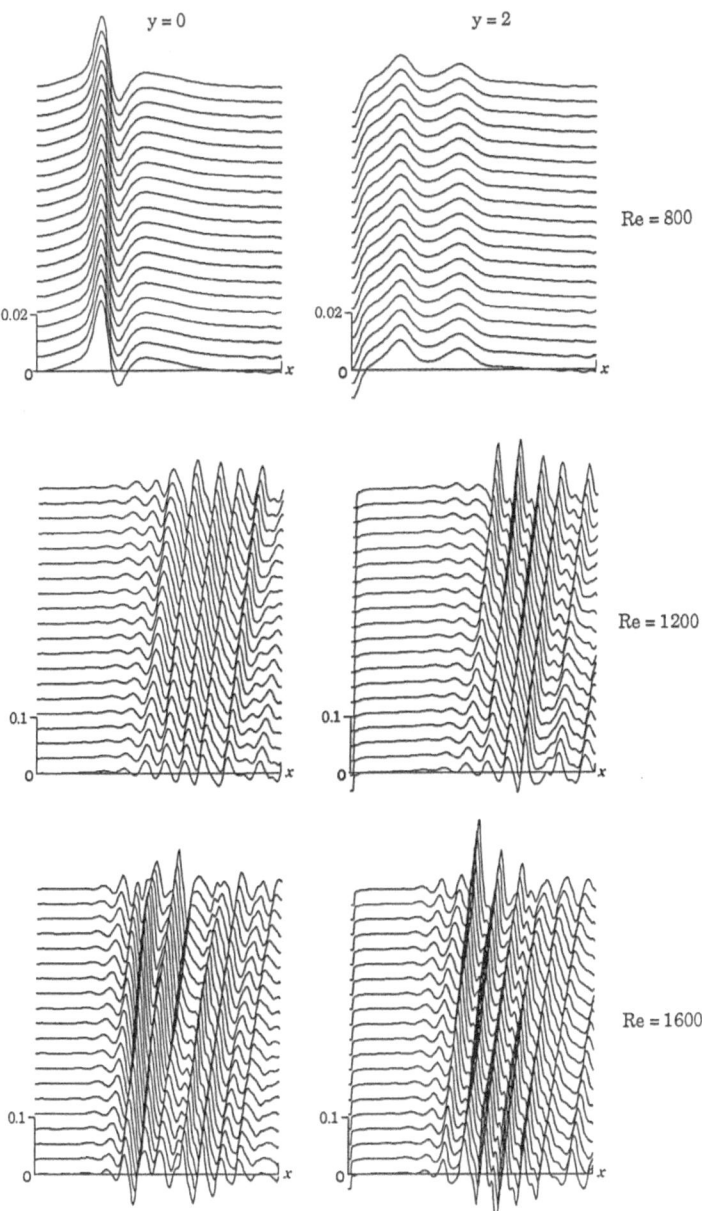

Fig. 5. Time variation of the pressure gradient in the x-direction ($0 \leq x \leq$ 40) for t = 1000 to 1020.

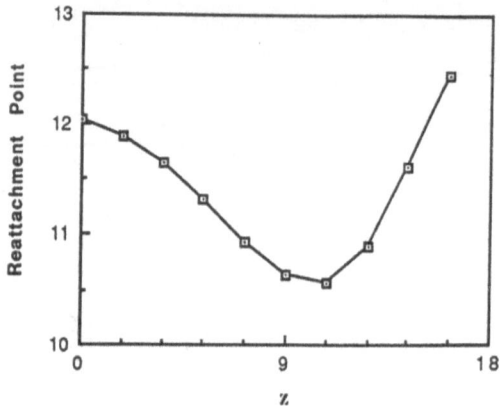

Fig. 6. X coordinate of the reattachment point of this stationary vortex behind the step as a function of z in the case of Re = 800.

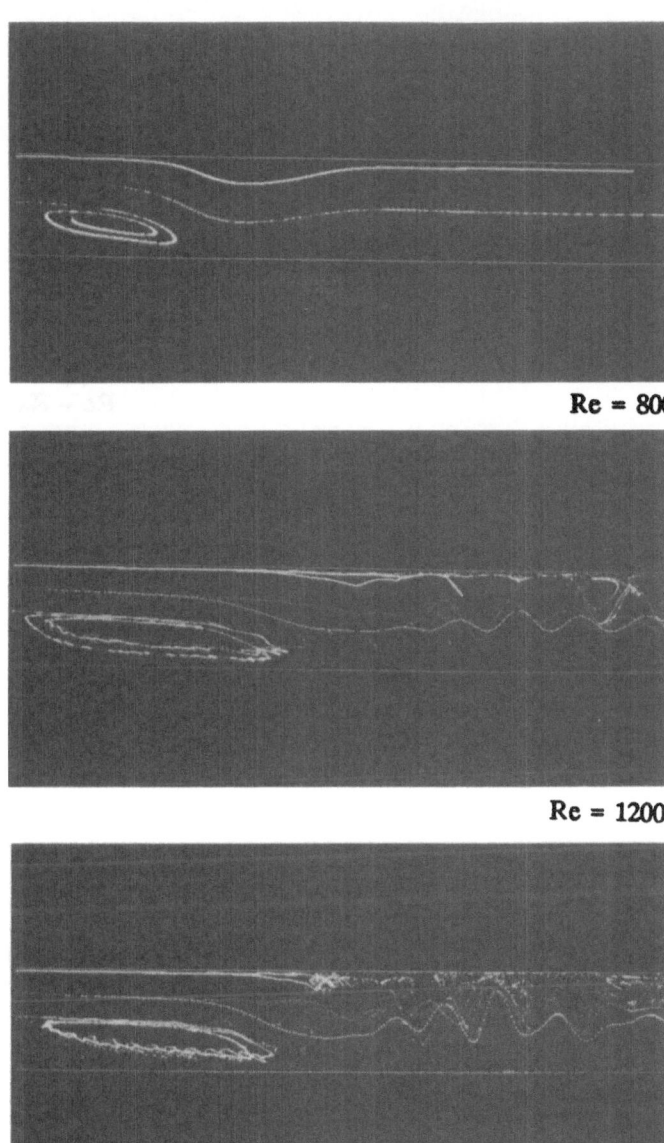

Re = 800

Re = 1200

Re = 1600

Fig. 7a. Side view of the streak lines at t = 1000. (Color art for this figure may be seen in the color insert.)

24

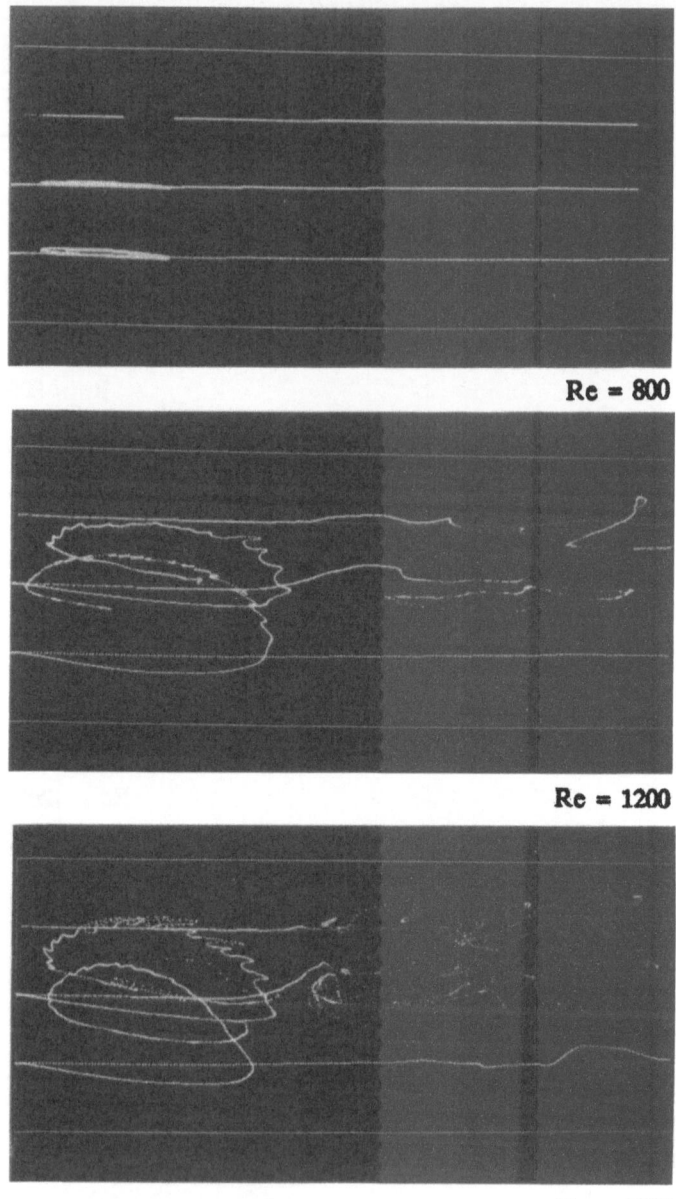

Re = 800

Re = 1200

Re = 1600

Fig. 7b. Top view of the streak lines at t = 1000. (Color art for this figure may be seen in the color insert.)

System Solutions for Visualization: A Case Study

Kohei Kumazawa *   *Christopher Eoyang* *

Abstract. *The increasingly important role of scientific visualization to the processes of analysis, simulation, and development has resulted in a number of widely different system configurations to compute and display data in a graphical form. In this paper we will compare a number of simulations and their visualizations using a well-known public-domain code, DYNA-3D, on three systems configurations: 1) supercomputer to graphics workstation; 2) superminicomputer to graphics workstation; and 3) graphics superworkstation. In particular we will emphasize the importance of communications speed to the time-to-solution in cases where visualization is a necessary part of the simulation process.*

1. Introduction

The widespread use of supercomputers and other high-volume sources of scientific data like satellites and seismological surveys has provided a strong impetus for research in methods for scientific visualization, or the conversion of numeric data into two- or three-dimensional graphical representations using color and geometry to represent data values. These visualizations are frequently animated to show the time-variant characteristics of the data. Visualization is rapidly becoming an essential part of the process of scientific discovery, enabling researchers to explore more of their data with greater efficiency and increased comprehension.

The process of scientific visualization poses problems which go beyond the capabilities of most of today's systems. Visualization (involving computation, rendering, and display) requires at least two machines in most cases, thereby necessitating some sort of data transfer between the two. With rare exceptions, a great deal of data must be transferred, especially in animated visualizations, for which the worst case (frame data is sent through the net for display) may involve up to 30 full-color 1000 x 1000 frames per second, requiring a bandwidth of almost 1 Gb per second. Present system configurations for visualization usually employ a high-speed computational engine (most often a supercomputer or a superminicomputer) for the computation, and a graphics workstation for the display. Some configurations, like those in use at the National Center for Supercomputing Applications (NCSA), use a multi-processing minisupercomputer (an Alliant FX-8) to perform high speed rendering for subsequent display on a high-resolution frame buffer. Most current networks, however, have nowhere near the capacity to handle the sheer volume of data

*Institute for Supercomputing Research, Recruit Co., Ltd.
8F Recruit KachiDoki Bldg, 2-11 Kachidoki, Chuo-ku, Tokyo 104 JAPAN

25

required by visualization, and the time-to-solution on these systems exhibits an inordinate dependence on network transfer speeds, regardless of the performance of the component machines.

Recently, however, systems like the Ardent/Kubota TITAN and Stellar GS1000, combining high speed vector/scalar computational engines with high-performance graphics hardware, have become available at prices in the range of upper-end engineering workstations. These graphics superworkstations have made it possible to compute, render, and display on a single machine, thus circumventing the network transfer problem.

In this paper we would like to explore three case studies in visualization using a number of system configurations loosely representative of those in use around the world. In particular, we are interested in comparing the time-to-solution performance of two conventional systems with one of the new graphics superworkstations, to give an idea of the viability of the latter. For our case study, we have chosen problems from the area of crash analysis, since it is one of the major areas in large-scale computing which benefit from visualization techniques (in this case, to determine the validity of a certain simulation run).

2. Background Description

Our present system network allowed us a number of different possible configurations to do the visualizations. Below, we give a short explanation of the three configurations and component hardware used in this paper.

2.1 System Configurations

System 1: CRAY-9600-IRIS
A Cray X-MP/216 with a Hitachi HITAC 8690 as a front-end connected to a Silicon Graphics IRIS 3130 graphics workstation by a 9600 bps line.

System 2: VAX-Ether-IRIS
A Digital Equipment VAX 8810 connected to a Silicon Graphics IRIS 3130 graphics workstation over a 10 Mb/sec (peak) Ethernet connection. A total of 15 other machines (minisupercomputers, array processors, and workstations) are also connected to this network.

System 3: TITAN
An Ardent-Kubota TITAN/2 with 2 processors used for both computation and rendering display. No network connection was necessary in any of the problems.

2.2 Hardware description

2.2.1 Cray X-MP/216
Our Cray X-MP/216 (in Tokyo) runs COS version 1.16 and the CFT 1.15 BF3 FORTRAN compiler. The two X-MP processors have a clock cycle time of 8.5 ns, and share a main memory of 16 MWords (128 MBytes). The machine was installed in 1985, and has a theoretical peak speed of 470 MFLOPS.

The X-MP/216 uses the Hitachi HITAC 8690 as the front-end machine. The front-end to X-MP connection is a 1.5 Mb/sec. HYPERchannel link over a dedicated digital line which has a peak transfer rate of 1.5 Mb/sec.

2.2.2 VAX 8810
The VAX 8810 superminicomputer from Digital Equipment runs VAX/VMS V4.7 with VAX FORTRAN V4.8–275 and VAX LINKER V04-00.

2.2.3 TITAN/2
The Titan/2, a two processor graphics superworkstation was designed by Ardent Computer and manufactured by Kubota Computer. This two-processor system, which has a theoretical peak speed of 32 MFLOPS and 32 MIPS, is one of a new class of machines aimed at coupling high computational performance (minisupercomputer class) with the graphics performance of a high-end graphics workstation. Ideally, this multi-functional machine would be capable of doing real-time simulation, i.e. simultaneous computation, rendering and display of a simulation in real time. The TITAN was used for both its computational and graphics capabilities in this paper.

2.2.4 Silicon Graphics IRIS 3130
The Silicon Graphics IRIS 3130 is a engineering/graphics workstation which is most frequently used for its high-end graphics capabilities and for displaying computational results in a graphical form. The machine runs UNIX System V3.5 Rev. 1. In this paper it was used with the PATRAN graphics post-processor to display the results of the simulations.

2.3 Problem Description
Each of the problems involves computation of the entire crash simulation using the DYNA-3D code and the subsequent visualization on a graphics workstation of 20 sample frames (to determine the validity of the simulation computed). In two of the system configurations (#1 and #2), this involves data transfer over a network between different machines.

2.3.1 DYNA-3D

DYNA-3D is an explicit three-dimensional finite element code for analyzing the large deformation dynamic response of inelastic solids and structures, originally developed in 1976 at Los Alamos National Laboratory, and subsequently revised in 1979, 1981, and 1982 [1]. This well-known code is widely used for doing three-dimensional crash and impact analysis. DYNA-3D includes a contact-impact algorithm which permits gaps and sliding along material interfaces with friction. Equations of motion are integrated in time by the central difference scheme, and the code includes twenty-five material models and eleven equations of state to cover a wide range of material behavior. The code is highly vectorized for optimal performance on vector processing supercomputers.

DYNA-3D runs on VAX/VMS, IBM, UNIX and COS operating systems, and thus it was possible to run this code on a number of our systems. The VAX/VMS version of the DYNA-3D code was ported to execute on the TITAN by the principal author, who was also responsible for writing the graphics routines necessary to display the DYNA-3D output on the TITAN.

All computational results presented here include all I/O and system time, in addition to CPU time, since we are primarily interested in the length of an actual simulation run.

2.3.2 Visualization

With the Cray and VAX 8810, visualization of DYNA-3D output data was achieved by using the PATRAN graphics post-processing package on the Silicon Graphics IRIS 3130 workstation. Execution on the TITAN required writing graphics routines in Doré, the high-level graphics language on the TITAN, in order to render and display the data in a manner similar to the PATRAN output.

Display graphics routines on the TITAN were done using independent polygons because of implementation considerations. In a more thorough implementation the TITAN's polygon mesh objects would be used to group polygons together and therefore the rendering and display times for the TITAN would be considerably faster than our results here.

In all fairness, using the PATRAN package almost certainly contributed to the relatively slow performance of the IRIS 3130 when compared to the TITAN. PATRAN is a full-featured general purpose code which covers much more than graphics display, and the results we obtained probably include a good deal of overhead not associated with the display routines.

Screen snapshots of the three visualizations are shown in Figures 1–3.

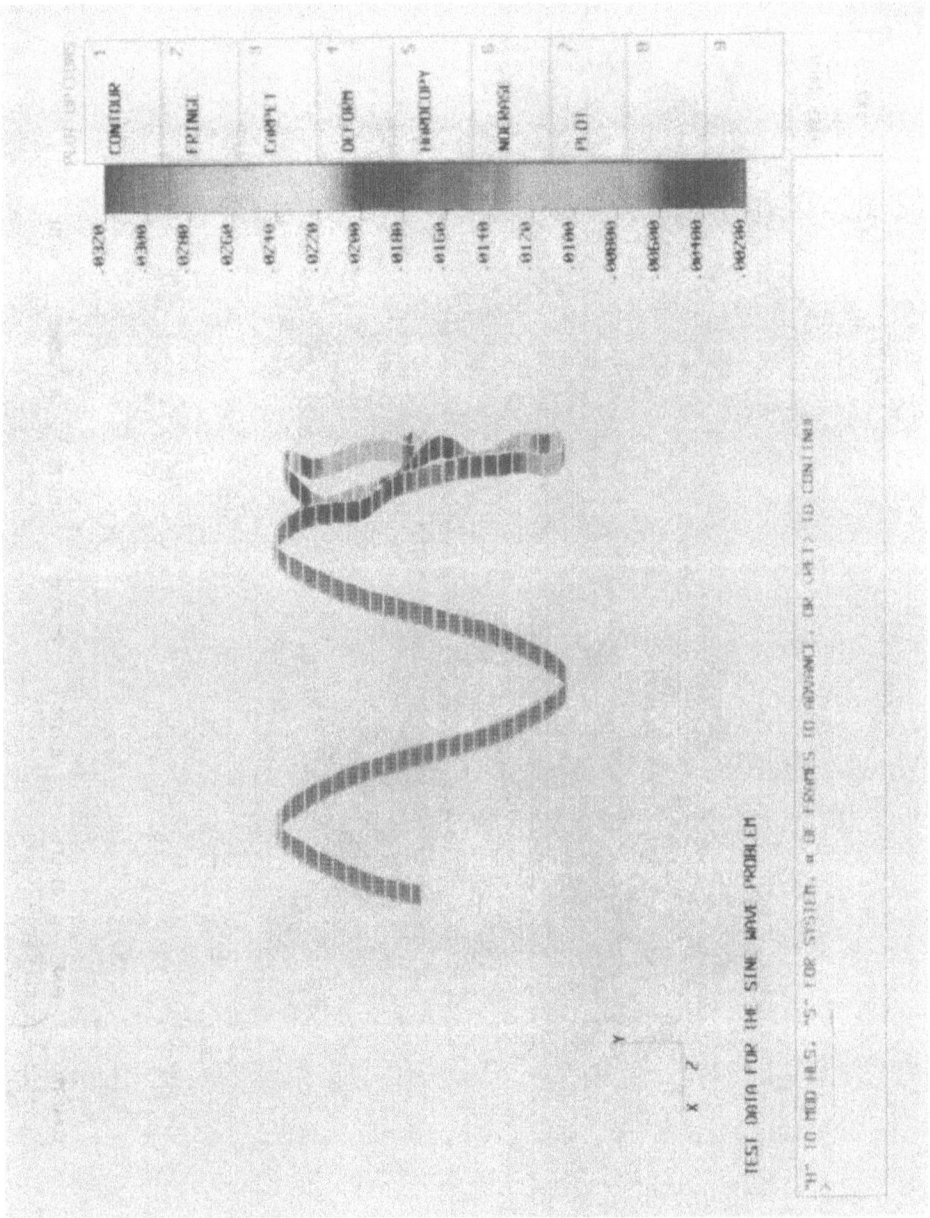

FIGURE 1. Sine wave spring. (Color art for this figure may be seen in the color insert.)

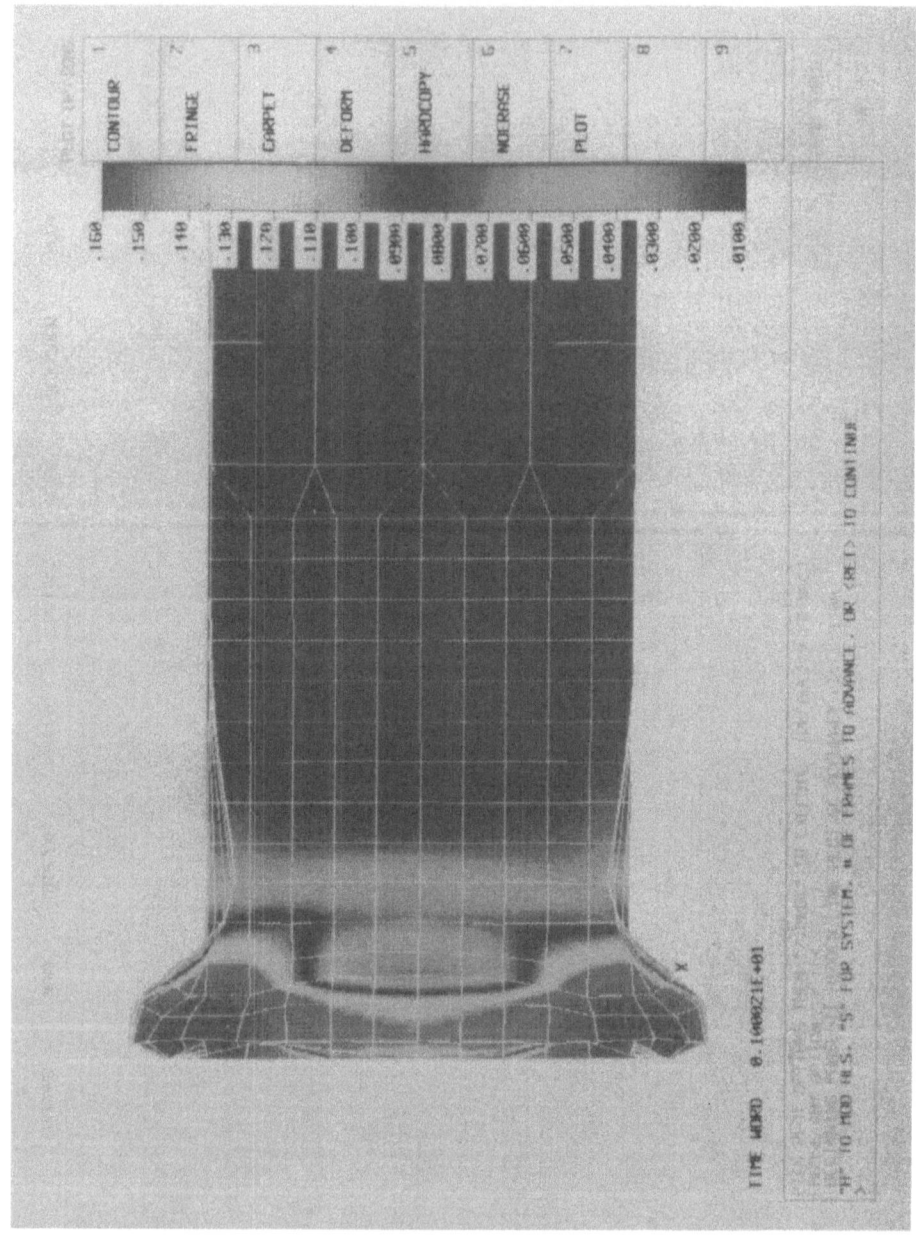

FIGURE 2. Rectangular tube. (Color art for this figure may be seen in the color insert.)

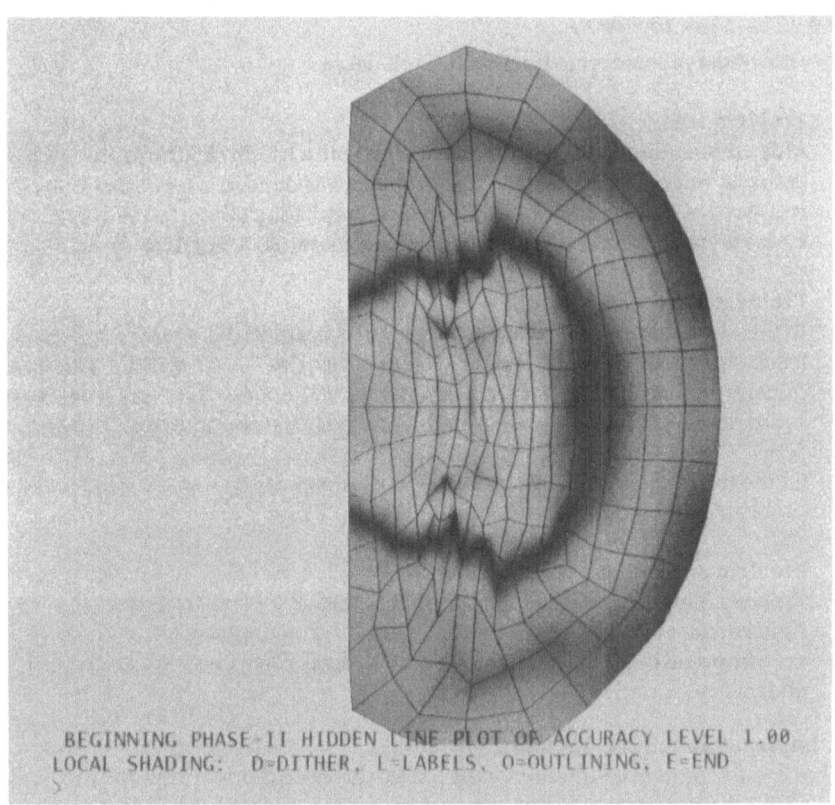

FIGURE 3. Circular plate. (Color art for this figure may be seen in the color insert.)

2.3.3 Three test problems

The three test problems are briefly outlined below:

Problem 1: Sine wave spring

This problem simulated a sine wave spring with a length of 50 mm hitting a rigid wall at a velocity of 50 m/sec. This problem was done in three-dimensions and consisted of 120 elements, 243 nodes, with a simulation time of 0.05 seconds. 4807 cycles were computed. Each cycle frame contains 88 KB of data.

Problem 2: Rectangular tube

In this simulation, a 160 mm long rectangular beam with a square cross-section (each side is 32 mm) hits a rigid wall at a velocity of 100 m/sec. The three-dimensional simulation has 889 elements and 872 nodes. The simulation covers 1 msec and 6807 cycles are computed. Each cycle frame contains 492 KB of data.

Of the three, this problem is perhaps most representative of the typical sort of problem one would be simulating on DYNA-3D.

Problem 3: Circular plate

Problem 3 involves a circular plate with a diameter of 150 mm crashing into a rigid barrier at 970 m/sec. 209 elements and 192 nodes are used, and 130 cycles are computed in the 0.05 msec of simulation time. Each frame consists of 111 KB of data.

3. Implementation and Results

The time-to-solution results are shown for the three problems in table form in Figure 4 and in the form of stacked-bar graphs in Figures 5, 6, and 7. Hypothetical results

	Computation	Communication	Display	TOTAL	Relative
Problem 1		(88 KB/Frame)			
CRAY/IRIS	68.2594	2300	280	2648	2.16
VAX/IRIS	951.786	308	280	1540	1.26
TITAN	1182.522	0	44	1227	1.00
Problem 2		(492 KB/Frame)			
CRAY/IRIS	338.143	8600	660	9598	2.05
VAX/IRIS	7378.91	640	660	8679	1.85
TITAN	4627.22	0	50	4677	1.00
Problem 3		(111 KB/Frame)			
CRAY/IRIS	2.197	2260	300	2562	11.24
VAX/IRIS	174.98	270	300	745	3.27
TITAN	182	0	46	228	1.00

FIGURE 4. Results (in seconds).

for a Cray – IRIS system configuration connected via Ethernet are also shown.

Though in some cases it is possible to overlap (pipeline) execution of the three operations and thus decrease total solution time, usually this cannot be done automatically due to operating system differences and/or the need for constant operator intervention. In some all-UNIX nets such as those using NFS and Cray's UNICOS, it may be possible to use pipes to efficiently route data and thereby overlap computation, communication, and display. In most cases this is still not possible, although we expect this capability will be available on the majority of future systems. We have thus opted to use the worst-case scenario, where each process begins only after the preceding process is completely finished.

In all cases the graphics performance of the TITAN is significantly better than the IRIS, due in part to the overhead incurred by the PATRAN post-processing package (see note in 3.3.2). This advantage in display performance has relatively minimal effect on the final results, however. We shall briefly discuss the results of each problem below.

3.1 Problem 1
Despite the Cray's nearly 20-to-1 advantage in computational speed over the TITAN, the vast amount of time spent doing data communications allows the time-to-solution on the TITAN to be twice as fast as on the Cray–IRIS configuration. The widely used VAX-IRIS setup is only slightly worse than the TITAN, due to the advantage in computational performance on the VAX and the relatively small amount of time spent doing data transfer.

If the Cray–IRIS configuration were to be connected via Ethernet instead of 9600 baud line, the performance is significantly better than the TITAN (almost twice as fast, but communication still accounts for almost half of the total time).

3.2 Problem 2
In the rectangular tube problem, the CRAY-9600-IRIS system is once again dominated by the slow transfer speed, and the TITAN is the fastest overall system. The VAX-Ether-IRIS system is slightly faster than the Cray despite the slow computational speed.

As might be expected, the CRAY-Ether-IRIS system would be close to five times faster than on the TITAN because of faster execution on the Cray. It is important to note the nice balance between computation, communication, and display with the Cray-Ether-IRIS system: if these processes were overlapped there is a good chance that the overall time could be substantially reduced because no single process dominates the execution time.

3.3 Problem 3
In the third problem, the circular plate, the communication cost of using the CRAY-

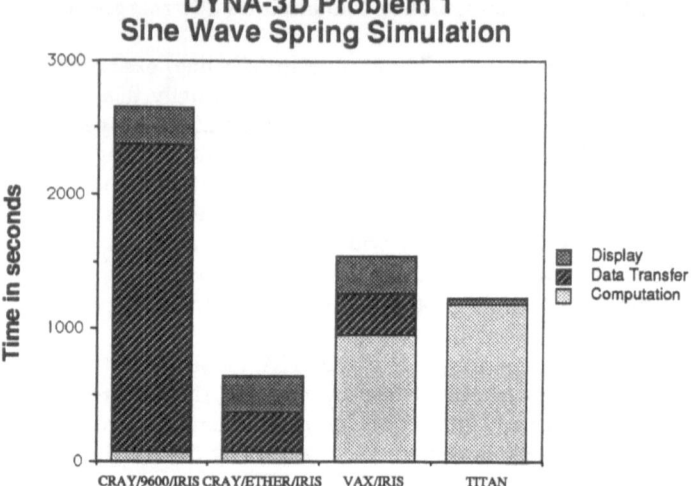

FIGURE 5. DYNA-3D: Problem 1--Sine wave spring simulation.

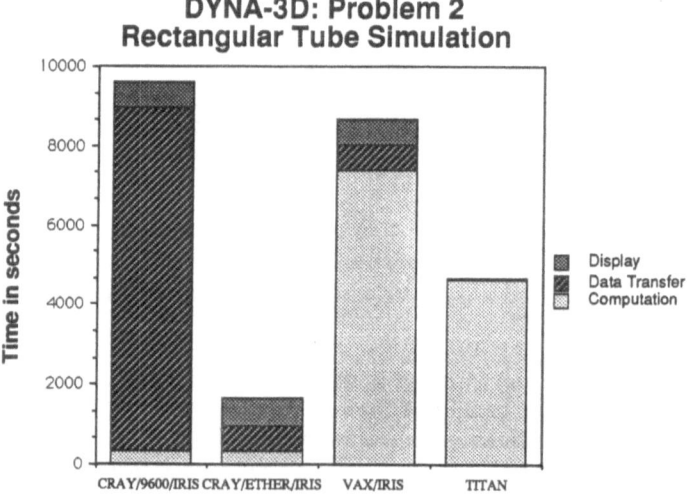

FIGURE 6. DYNA-3D: Problem 2--Rectangular tube simulation.

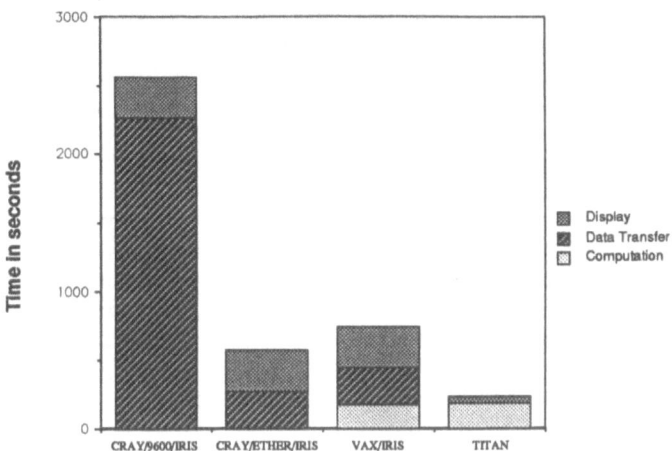

FIGURE 7. DYNA-3D: Problem 3--Circular plate simulation.

9600-IRIS system is three orders of magnitude greater than the computation cost, and overall, the TITAN configuration is 11 times faster than the Cray, and 3 times faster than the VAX.

Even the CRAY-Ether-IRIS system, with a dramatic increase in communication speed, is more than a factor of two slower than the TITAN system. When real time animation was attempted on the TITAN, the entire simulation was calculated and rendered in 192 seconds, 34 seconds faster than the composite times, an improvement of 15%, or almost 3 times the performance of a Cray-Ether-IRIS system.

4. Problem Analysis: Time-to-solution
Our results clearly show that unless network speeds are balanced with computational speeds, the time-to-solution will depend more on communications than computational performance. The results given above could very well be reached with simple common-sense. They help to dispel the notion that increased computational power alone will result in increased productivity and throughput: if an equal investment is not made in networking, the effect of any increase in computational speed will be drastically reduced.

The time to compute a solution is not the time-to-solution. Just as a product must be delivered before it becomes useful, so must a user receive results in some sort of comprehensible form before one can say that the problem has been completed.

Since the dawn of the computer, far more energy has been spent in making computers go faster than in making it possible for them to communicate faster, even though data transfer rates affect the time-to-solution just as much as the actual computation. Communications, in many ways, have been relegated to secondary status behind computation by people who erroneously believe that connectivity is a synonym for communications.

The X-MP is no longer a state-of-the-art computer. Newer models, with peak speeds an order of magnitude greater, capable of actually sustaining computational rates of over 1 GFLOPS in some cases are just starting to become available. Furthermore, corresponding increases in main and extended memory capacities have made larger and larger data sets possible, with up to 2 GB of main memory available. With this surge in computational power and data capacities, the computation–communication imbalance will only worsen. Users at many installations, including ours, still access supercomputers via 9600 baud lines, which is becoming ridiculous in light of the enormous amount of data being transferred, the tremendous expense of using a supercomputer, and the constraints this imposes on the user.

As machines like the TITAN become faster and more common, it will be more feasible to handle many problems now computed over a network on a single machine, thus circumventing the network problem as well as the need to share supercomputer time with other users. While this will solve some network problems by avoiding networks all together, this will not solve the need to share one's results with one's colleagues and other researchers.

5. Conclusion
We have taken three sample simulations/visualizations from crash analysis to do a case study on the efficiencies of different system configurations. Our results show that in many of today's configurations, communications present the greatest bottleneck, and that if low-speed networking is used for data transfer, graphics superworkstations may in fact be faster than systems with supercomputers in terms of the total time-to-solution.

A chain is only as strong as its weakest link. Compute-intensive technologies like simulation, visualization, image processing have encouraged massive efforts to develop faster computers with more processors, faster devices, and more capacity. But in networking, where current data transfer rates already lag far behind present computational capacity, a corresponding awareness of similar needs for high-speed communications is much less prevalent. As C. Gordon Bell has argued, "Access to information has never been more important than it is today, and the ability to fully exploit information resources — be they individual researchers, research teams, databases, or supercomputers — will determine how competitive any group or nation is" [2].

6.0 References

[1] DYNA-3D Users Manual

[2] C. Gordon Bell, "A National Research Network", IEEE *Spectrum*, February 1988, pp. 54-57.

Part 2

Visualization Hardware/Performance

A GENERAL APPROACH TO NONLINEAR DYNAMIC ANALYSIS
ON PARALLEL/VECTOR COMPUTERS

Robert E. Fulton, Professor

Kuo-Ning Chiang, Graduate Research Assistant

George W. Woodruff School of Mechanical Engineering

Georgia Institute of Technology

Atlanta, Georgia 30332

1.0 Abstract

The finite element method is widely used as a computational method to model physical system in various engineering problems. For detailed analyses of complex designs, structural models composed of several thousands of degree of freedom are no longer uncommon. The computer simulation of nonlinear dynamics response of structures, visualization and the implementation of parallel FEM systems on a high speed multiprocessor have received considerable attention in recent years (Ref. 1-3). The drivers of these activities included the reliable simulation of automotive, aircraft crash phenomena and increased performance of computers (Ref. 4-6). The development of general purpose finite element structural analysis computer programs have provided the capability to address a wide range of structures problems (Ref. 4). However, these software systems are severely limited for nonlinear transient analyses such as offshore structure, crashworthiness and aircraft design, etc. because of the available speed on current sequential computers. Parallel/Vector computers offer the promise for solution to complex problems in detail considered too time-consuming for todays sequential computers (Ref. 4, 7 and 8). To benefit from advances in parallel computers, software must be developed which take maximum advantage of parallel/vector processing features.

2.0 Introduction

The design of complex engineering systems such as advanced offshore platforms and car crash simulations requires continually increased levels of detail in supporting analyses. Such design activities require large order finite element models and excessive computation demands in both calculation speed and information management. Recent advances in parallel supercomputer technology provide the opportunity to upgrade current structural analysis capabilities. Most existing major structural analysis software systems were designed ten to twenty years ago and have been optimized for current sequential computers. Such systems often are not well structured to take maximum advantage of the recent and continuing revolution in parallel vector computing capabilities. These parallel vector computer architectures are not only occurring in the form of large supercomputers, but also are occuring for minicomputers and even engineering workstations. This paper discusses some recent advances in solving nonlinear structural dynamic problems and the development of parallel processing software for finite element systems. The parallel implementation criteria that influence the efficiency of an algorithm include the amount of computation versus the number of processors, the communication paths or overlap boundary length, waiting and synchronization delay, critical regions of algorithm that must be executed sequentially and the size of problem in relation to the number of processors used.   Finite element analysis and optimization of structures subjected static or dynamic forces typically require to the solution of large systems of linear equations, element and force generations. Many commercial finite element systems use a triangular decomposition of the system matrix in combination with a forward/backward substitution to solve the global equilibrium equations. For example in a static stress analysis it may take more than 50% of the total execution time (Ref. 12).   Recently the parallel matrix decomposition has been investigated by several researches (eg. Ref. 13-15) but the proposed solutions do not readily fit into the environment of multi-purpose finite element systems. One focal point of the present investigation is to develop and test a method for the parallel triangular decomposition of symmetric matrices for large scale finite element problems and the incorporation of these concepts into a production finite element system.

In the area of nonlinear dynamics, research is critically needed on the development and evaluation of parallel/vector methods for transient analysis of complex nonlinear structural problems. This study reports on an investigation of selected nonlinear dynamics algorithms appropriate for parallel/vector computers. Both implicit and explicit methods are considered and the merit of parallel/vector computations is investigated on four parallel computers, FLEX/32 shared/local memory multicomputer, CRAY X-MP/48 and IBM 3090-600E shared memory parallel/vector computers and Intel iPSC Hypercube local memory computer.

3.0 Relevant Progress in Parallel Hardware
Significant advances have taken place over the past year in parallel/vector supercomputer hardware which can provide directions relative to finite element computations and appropriate software. It is useful to classify the evolving supercomputer family into four categories, moderately parallel high speed system, massively parallel system, moderately parallel intermediate systems, and high performance parallel work stations. Table 1 gives a list of the parallel/vector computers in the four classes.

Table 1 shows that at one extreme are the high speed moderately parallel systems such as CRAY and IBM systems. At the other extreme are the massively parallel systems such as hypercube or butterfly systems. In between are various intermediate systems such as Alliant and FLEX/32 system, this latter group indicates the evolution of the established workstation market toward the new technology of parallel/vector processing; Table 1 gives data on speed and memory for these systems. CRAY has also recently announced its plans for the future (Table 2) which indicate a growing level of parallelism to go with its vector and large memory capability.

These results lead to several significant conclusions relative to the future of parallel processing.

1. The moderately parallel large scale computers are an established applications market with CRAY now selling 50% of its computers to industrial organizations. Furthermore, CRAY is moving steadily toward increasing numbers of parallel processors. The ETA and IBM activities make it a competitive market.

2. The massively parallel computer market (e.g. Hypercube, Butterfly, etc.) is not yet well established and usages are basically research oriented (some of it FEM oriented). Interest is still growing but the absence of general purpose FEM software for such machines inhibits their practical use. Nevertheless there is growing set of special purpose software; and some increased analysis capabilities are beginning to evolve.

3. The moderately parallel intermediate system market (e.g. Alliant, Convex) is still evolving and these systems are being used as less expensive alternatives to the CRAY or as CRAY frontends. Informal projections are that Alliant will soon move to a 16 or 32 processor capability.

4. The recent explosion on the scene (Table 1) of a large number of moderately parallel/vector based high performance workstations is an especially important event and is indication of the future. Both the Ardent and Raster systems have parallel and vector capability and Apollo and Raster have rated performance exceeding 100 MFLOPS. Such systems provide a parallel/vector graphics and computation capability on the engineering desk. These trends suggest that high performance parallel workstations will soon be highly competitive with the parallel intermediate systems such as Alliant and Convex. This would be similar to that which occurred when workstations began to take over much of the minicomputer market for sequential processors.

The results indicate the continuing evolution of the massively parallel computer market. This may occur through the acceptance of innovative minicomputer systems or by attached accelerators based on the hypercube or Butterfly configurations. The massively parallel approach for the high speed systems also appears to be steadily occuring through the continued growth of the CRAY type general purpose computers, with CRAY planning at least 64 processors in 3-4 years. IBM is also known to be studying massively parallel architectures. These results strongly indicated that the hardware base will evolve for large number of parallel/vector processors. Commercial finite element systems in 3-5 years will be needed which run effectively on 100-1000 processors.

4.0 Nonlinear Dynamics Equilibrium Formulation

The dynamic equilibrium balance equations for time step i can be expressed as

$$R_{int,i} + R_i^I + R_i^D = R_i^E \tag{1}$$

The subscripts int, superscript I, D and E denote internal forces, inertia force, damping forces and external forces. The internal reaction force vectors are computed element by element then transformed from the local co-rotated element coordinate system (ref. 11) to the global reference system and assembled for all elements. The inertia force vector and the damping force vector are also in reality computed element-wise and assembled with use of the global matrices M and C. The incremental form of the governing equation may be written as (ref. 11)

$$M_i\Delta\ddot{u} + C_i\Delta\dot{u} + K_{I,i}\Delta u = R_i^E - (R_{i-1}^I + R_{i-1}^D + R_{int,i-1}) \tag{2}$$

M_i is the mass matrix, C_i the incremental damping matrix and $K_{I,i}$ the incremental stiffness matrix. $K_{I,i}$ may be taken as the tangential element stiffness K_T, assembled to a global system matrix, or, the K_I may be used for several consecutive steps. $\Delta u_i, \Delta\dot{u}_i, \Delta\ddot{u}_i$ are the incremental displacement, velocity, and acceleration vectors. The Δ denotes finite but "small" increments corresponding to the difference between the two states considered. Equation (2) is used in connection with load increments and equilibrium iterations.

4.1 Implicit Time Integration Algorithm

A general incremental-iterative scheme combined with the Newmark family of implicit time integration operators is used to solve the dynamic equations. The basic assumptions of the Newmark operators are

$$\dot{u}_{i+1} = \dot{u}_i + (1-\gamma)\ddot{u}_i h + \gamma\ddot{u}_{i+1}h \tag{3}$$

$$u_{i+1} = u_i + \dot{u}_i h + (\frac{1}{2}-\beta)\ddot{u}_i h^2 + \beta\ddot{u}_{i+1}h^2 \tag{4}$$

Here h is the incremental time step

$$h = \Delta t = t_{i+1} - t_i$$

which may vary throughout the analysis. From equations (3) and (4) the corresponding increments of velocities and accelerations are

$$\Delta\dot{u}_{i+1} = \frac{\gamma}{\beta h}\Delta u_{i+1} - \frac{\gamma}{\beta}\dot{u}_i - (\frac{\gamma}{2\beta}-1)h\ddot{u}_i \tag{5}$$

$$\Delta\ddot{u}_{i+1} = \frac{1}{\beta h^2}\Delta u_{i+1} - \frac{1}{\beta h}\dot{u}_i - \frac{1}{2\beta}\ddot{u}_i \tag{6}$$

Substitution equations (5) and (6) into equation (2) yields

$$\left[M\frac{1}{\beta h^2} + C\frac{\gamma}{\beta h} + K\right]\Delta u_{i+1} = R_{i+1}^E - (R_i^I + R_i^D + R_{int,i})$$
$$+ M\left[\frac{1}{\beta h}\dot{u}_i + \frac{1}{2\beta}\ddot{u}_i\right] + C\left[\frac{\gamma}{\beta}\dot{u}_i + (\frac{\gamma}{2\beta}-1)h\ddot{u}_i\right] \tag{7}$$

then the incremental - corrective equations become

$$\hat{K}\Delta u_{i+1} = \Delta\hat{R}_{i+1} \tag{8}$$

where $\hat{K}$ is the effective stiffness matrix, $\hat{R}$ is the effective nonlinear force vector.

and $\Delta\dot{u}_{i+1}, \Delta\ddot{u}_{i+1}$ can be obtained from equation (5), (6). The corresponding total vectors are

$$u_{i+1} = u_i + \Delta u_{i+1}$$

$$\dot{u}_{i+1} = \dot{u}_i + \Delta \dot{u}_{i+1} \qquad\qquad (9)$$

$$\ddot{u}_{i+1} = \ddot{u}_i + \Delta \ddot{u}_{i+1}$$

If the vectors $u_{i+1}, \dot{u}_{i+1}$ and $\ddot{u}_{i+1}$ (equation 9) are substituted into the dynamic equilibrium equation (1), it will be found for nonlinear cases that this equation is not completely satisfied, and equilibrium iterations must be carried out.

4.2 CPU and Storage

The major time-consuming steps of implicit method are (1) nonlinear force vectors generation, (2) input/output, and (3) matrix decomposition. The mass matrix, stiffness matrix, viscous damping matrix, internal force vectors can be computed element by element and assembled into the global matrices M, K, C_V and R_{int}. These computational tasks are independent and highly parallelizable and can be assigned across processors (Ref. 4). The Cholesky decomposition step is the most time-consuming calculation for large size finite element problems, and this method has shown good speedups through parallel and vector processing, especially when the half-bandwidth is sufficiently large (ref. 7, 10).

For a large structural system the effective stiffness matrix cannot be accommodated in core storage (shared memory), and must be segmented into blocks. These blocks must then be stored temporary on low speed auxiliary storage, usually as disk files, where the access time is relatively high (Ref. 4). Shared/Local memory can take advantage of the local memory available on each processor, Data blocks are stored in local core memory by use of standard FORTRAN 77 calculations such as Local Variables 1 to n equal to Share Variables 1 to n, these data mapping CPU times are much less than the standard I/O operation. Therefore, the use of local memory as a secondary storage unit is needed for parallel finite element implementations.

The equation (8) may take the form as A X = B, where the effective stiffness matrix $A = L U = L D^{-1} L^T$. Here the matrix L is a lower triangular matrix and D is a diagonal matrix, the decomposition is carried out in three step:

$$l_{ij} = a_{ij} - \sum_{n=1}^{j-1} l_{ni} l_{nj} \qquad i < j$$

$$l_{ij} = l_{ij}\, d_{ii}^{-1} \qquad\qquad (10)$$

$$d_{jj}^{-1} = 1 / (a_{jj} - \sum_{n=1}^{j-1} l_{nj} l_{nj} d_{nn}) \qquad \text{for all } j$$

The matrix L has the same profile as A (skyline form) and is therefore stored in the same location. The diagonal matrix D^{-1} is stored in an array residing in core. The stiffness matrix A is divided into several blocks, each consisting of several columns and A is triangulized block by block (Ref. 4). Some typical results of this algorithm are shown on the following section.

5.0 Parallel/Vector Matrix decomposition

Concurrency is achieved by mapping matrix columns across processors (Figure 1). One processor computes all terms l_{ij} with $i = 1$ to $j-1$ and the diagonal term d_{ij}^{-1} in one particular column j using equation (10). Concurrently the other processors factor the neighboring columns $j+1$, $j+2$, etc. However these tasks are not completely independent (Ref 17). Software flags and lock variables are used to synchronize the decomposition procedure when the tasks reach the critical region at the bottom of the column. In the parallel/vector approach used here parallelization is done on the outer loop and vectorization is applied to the inner loop of equation 10, which correspond to dot product of two matrix columns, efficiently vectorized dot routines are readily available for many computers and the approach taken leads easily to vector computing on multiple processors.

The parallel decomposition procedure was initially coded and tested on a FLEX/32 multicomputer (Figure 2). The FLEX/32 is a multiple instruction multiple data machine with a shared memory capability and served effectively as a test bed for the subsequent parallel supercomputer implementation. Application of the decomposition as a part of a finite element solution procedure usually implies that the matrix is not fully populated but has nonzero terms clustered around the diagonal. Skyline matrices were, therefore, used to investigate program performance. Figure 3 shows the speedup obtained for 500 D.O.F system using up to six processors. Here speedup is defined as the ratio of sequential to parallel execution time of decomposition module. Figure 3 shows that when the mean half-bandwidth exceeds 50, the speedup remains close to linear optimum, and processor utilization depends strongly on the matrix mean half-bandwidth; however, for constant mean half-bandwidth the variation of system size has only moderate effect on program performance. The theoretical speedup limit is taken to be equal to the number of processors because the total decomposition task is considered to be executed in parallel. The same program has been tested on CRAY X-MP/48 at Pittsburgh Supercomputer Center for investigating vectorization speedup (Figure 4) and also tested on IBM 3090-600E at Cornell Supercomputer Center for investigating parallel/vector performance (Figure 5, 6). In the parallel Cholesky decomposition, each processor is responsible for decomposing the whole column of the system matrix; therefore no vector length penalty is introduced (Figure 5, 6). Vectorization performance of this matrix decomposition has been optimized for CRAY X-MP/48 that is the reason why CRAY X-MP/48 shown better vectorization results than IBM 3090-600E.

The numerical experiments on the FLEX/32, CRAY X-MP/48 and IBM 3090- 600E gave excellent program performance for relatively small system matrices. For complex structures typical finite element models lead to system matrices which are larger in both size and bandwidth than the test problems considered here. The results of Figure 3-6 suggested good parallel/vector performance should be expected for such applications.

6.0. Application to an FEM Crash Analysis

The parallel methods have been incorporated in a nonlinear finite element program and applied to the analysis of a structure subjected to crash type condition. The finite element system used is FENRIS, a large scale, general purpose program for nonlinear finite element analysis. FENRIS was developed as a project between the

Norwegian Institute of technology (NTH), the Society for Industrial and Technical Research (SINTEF) and the Norwegian VERITAS. Modification to FENRIS included implementation on a parallel operating system and replacement of selected sequential code by parallel code. The resulting parallel system denoted FENRIS parallel version one (FENRIS/P1, Figure 7) with a parallel $LD^{-1}L^T$ had been developed and installed on FLEX/32 MMOS (Multitasking Multicomputing Operating System) at Georgia Tech CAD/CAE Laboratory in 1987. Speedup of the $LD^{-1}L^T$ decomposition (compare with the sequential FENRIS $LD^{-1}L^T$ decomposition) are very encouraging. The improvements are achieved by refined parallel computation strategies, and a machine independent waiting control routines used to reduce waiting time in the matrix decomposition.

FENRIS/P1 was applied to a crash test problem shown at Figure 8, which is an automobile front end S-shape torque box beam subjected to impact. This finite element model comprises 180 quadrilateral shell elements with drilling freedoms and mounting to 1140 D.O.F., the maximum half-bandwidth is equal to 42. The impact phenomenon was computed on the basis of elastic-perfectly plastic constitutive material (A-36 Steel). Dynamic computations were performed via a matrix solution of Newmark's unconditional stable scheme ($\alpha = 0.5$, $\beta = 0.25$), with a time increment equal to 0.0001 second and a true Newton-Raphson iterative scheme. A fine mesh was not used because of CPU time and disk space limitations on the FLEX/32. The typical results showing the improved computation speedups are given in Figure 9. The results show computation speedup for $LD^{-1}L^T$ of up to 6.32 for seven processors and indicate that significant speedups can be achieved through the use of many processors for appropriate finite element problem. From the above observation, the $LD^{-1}L^T$ algorithm shows good promise for parallel/vector computation and indicates that a Cholesky based finite element system such as FENRIS appears well suited for parallel/vector implementation. Figure 11 shows the correlation between experimental results and the parallel computation strategies.

7.0 Parallel Explicit Time Integration

An alternate timewise integration of equation (2) can also be done through an explicit central difference approximation for u_j and u_{j+1}, writing equation (2) at t_j leads to

$$M_i \Delta \ddot{u} = R_i^E - (R_{i-1}^I + R_{i-1}^D + R_{int,i-1}) - C_i \Delta \dot{u} - K_{i,i} \Delta u \tag{11}$$

where

$$
\begin{aligned}
u_{j+1} &= \ddot{u}_j h^2 + 2u_j - u_{j-1} \\
\dot{u}_{j+1} &= 2\ddot{u}_j h + \dot{u}_{j-1}
\end{aligned}
\tag{12}
$$

Explicit methods such as equation (11) lend themselves to assigning equations for a select set of nodal variables to specific processors (Ref. 16). The solution procedure has thus been parallelized according to physical regions, and might be denoted "physically parallel" (or subdomain parallel). On the other hand, if the solution procedure is based on a Cholesky type decomposition. The solution has been parallelized mathematically and might be denoted as "algorithmically parallel". Thus the integration approach, system matrices, number of available processors, and other features may significantly affect the solution approach most appropriate for a problem and the attendant benefits of parallelism.

To investigate substructural parallelism consider the nonlinear dynamic response of a simply supported shallow arch subjected to a step pulse. A parallel solution to the arch was carried out at Georgia Tech on the FLEX/32 and similar results were obtained at General Motors on the Intel iPSC Hypercube.

The governing equations for the arch are shown in Figure 11 as

$$E I \left[\frac{\partial^4 w}{\partial x^4} - \frac{d^4 w_o}{dx^4} \right] - N \frac{\partial^2 w}{\partial x^2} + \rho A \frac{\partial^2 w}{\partial t^2} = -p(x,t) \tag{13}$$

where

$$N = \frac{E A}{2} \left[(\frac{\partial w}{\partial x})^2 - (\frac{dw_o}{dx})^2 \right]$$

in which:

$w(x,t)$ = transverse displacement of the middle surface of the arch
$w_o(x)$ = initial shape of the arch
ρ = mass density of the arch material
A = cross-sectional area of the arch
p = distributed force per unit length
I = Moment of Inertia
E = Young's Modulus

In equation (13) central finite differences are used to approximate spacial derivatives and central differences are used for time integration similar to that discussed in equation (2-9). If the number of D.O.F is a multiple of the number of processors n, then equation (13) can be mapped on a parallel computer with computation distributed equally across processors. If the number of D.O.F is not a multiple of n, some imbalance in processor workload will occur, however, for a large D.O.F. problem this workload imbalance negligible. An executive may be designed in some cases to partition subordinate tasks among processors, help balance process work load, and minimize idle processors and attendant overhead.

Integration of non-linear dynamic equations by the central difference method provides a direct approach for calculating the solution at the $j+1$ time step from information at j time step. Figure 12, 13 illustrate, for example, how dynamic equations of motion are mapped onto processors for both the FLEX/32 and the Intel iPSC. In this implementation each processor performs essentially the same task and the integration computations are interrupted once at the end of each time step for communication. Some minor imbalance in computation may occur for equations near the boundary due to minor differences in the equations there. This approach was implemented on the FLEX/32 using the shared memory common bus communication logic. Standard FORTRAN 77 was used for waiting control routine that is a big advantage of shared memory architecture from portability point of view, and Concurrent FORTRAN was used only for creating processors. The approach was also implemented on the Intel iPSC Hypercube using the interprocessor communication logic.

Standard FORTRAN 77 and INTEL Concurrent FORTRAN was used for creating processors, communications, and initializations.

Figure 14 shows typical results for the nonlinear dynamic response of an arch subjected to a step pulse. Figure 15 shows that the computation speedup for the FLEX/32 implementation as the number of degrees of freedom (D.O.F.) increases. This result was obtained using the shared memory communication. Similar results for the iPSC are shown in Figure 16. The somewhat better efficient of the FLEX/32 on the iPSC for the arch problem is due to the shared memory waiting control being faster than the local communication. Figure 15 shows that for 20 D.O.F. per processor a significant improvement in computation speedup occurs and the processor efficiency approaches 96% for six processors on the FLEX/32. On the other hand, with 10 D.O.F. per processor, the efficiency drops off since less computations are required on each processor. Furthermore for 20 D.O.F. per processor the overhead factors are relatively small for such factors as communication, lack of synchronization, and unequal distribution of computation tasks are relatively small. The 96% efficiency achieved with six processors provides a good indication of the potential benefits that can be expected from a moderate number of processors. The basic results also indicate that the benefits of parallel calculations drop off for those problems which require less extensive calculations and more intensive data communications. A comparison of the FLEX/32 and iPSC results illustrates that a shared/local memory machine such as the FLEX/32 architecture may be more efficient than a local connection computer like the iPSC for small physically parallel problems such as the arch.

8.0 Concluding Remarks

A study has been conducted evaluating the potential of parallel/vector computers to improve the computation capabilities for nonlinear finite element and finite difference analyses. Numerical methods typically used in FEM systems have been investigated on test problems with a view toward the steps which are computationally intensive and how these steps lend themselves to parallelism. Key computation steps investigated include matrix decomposition and dynamics algorithms. Results have also been obtained for an explicit time integration method where the mass matrix is diagonal.

The results indicate that the key computation steps can benefit significantly from parallel/vector computing. They also indicate that there are minimum thresholds in computation tasks for multiple processors to be effective. Below these thresholds data communication becomes a bottleneck and parallel computation efficiency significantly deteriorates. The results suggest that for problems with a large number of D.O.F and a moderate number of processors, good processor efficiency can be obtained. It could be shown that parallel/vector code can be readily incorporated in major finite element production code such as FENRIS.

The initial results are encouraging and suggest that more detailed studies should be carried out with all major computation steps of a FEM procedure being implemented in parallel.

Acknowledgements

Portion of this work were supported by the General Motor Research Laborato-

50

ries under contract E25-633 and in corporation with Norway Det Norske Veritas, and were conducted on the Cornell National Supercomputer Center and Pittsburgh Supercomputer Center Facilities.

REFERENCES

1. Fulton, R. E. "The Impact of Parallel Computing on Finite Element Computations". Presented at International Conference on Reliability of Methods for Engineering Analysis, Swansea, U.K. (9-11 July 1986).

2. Noor, A., Storaasli, O., and Fulton, R. "Finite Element Technology in the Future". Impact of New Computations on Computational Mechanics. (Noor and Pilkey, Editors), ASME Special Publication H00275 (November 1983) pp. 1 - 32.

3. Noor, A. K. and Atluri, S. N. "Advances and Trends in Computational Structural Mechanics", AIAA Journal, Vol. 25, No. 7, pp. 977-995, July 1987.

4. Fulton, R. E., and Chiang, K. N., "Computational Crash Dynamics Methods for Fifth Generation Supercomputers". GM Project E-25-633 Final Reports Phase III, Dec. 1988.

5. Argyris, J., Balmer, H. A., Doltsinis, J. St., and Kurz, A., "Computer Simulation of Crash Phenomena", International Journal of Numerical Methods in Engineering, Vol. 22, pp. 497-519, 1986.

6. Supercomputers: "The Proliferation Begins, Electronics", pp. 51-77, March 3, 1988.

7. Fulton, R. E., and Chiang, K. N., "Comparison of Shared Memory and Hypercube Architectures for Structural Dynamics", Third International Conference on Super-computing and Second World Supercomputer Exhibition, Proc. Vol. 1, pp. 418-426, May 15-20, 1988, Boston, Massachusetts.

8. Storaasli, O. O., and Bergan, P., "A Nonlinear Substructure Method for Concurrent Processing Computers", AIAA Publ. 86-0852, 1986.

9. Malone, J. G., "Automated Mesh Decomposition and Concurrent Finite Element Analysis for Hypercube Multiprocessor Computers", G.M. labs. Report, Engineering Mechanics Department, Warren, MI. May 18, 1987.

10. Chiang, K. N., and Fulton, R. E., "Nonlinear Dynamics Methods for Parallel Computers". ASCE Fifth Conference on Computing in Civil Engineering, Proc. pp. 144-158, March 29-31, 1988, Alexandria VA.

11. Nygard, M. K., "The Free Formulation for Nonlinear Finite Elements with Applications to Shells", Report No. 86-2, Division of Structural Mechanics, The Norwegian Institute of Technology, Norway, Dec. 1986.

12. Komzsik, L., "Parallel Decomposition Technique on a Double Level Storage System", International Conference on Vector and Parallel Processing, Loen, Norway, 1986.

13. Chen, S. S. and Dongarra, J. J., "Multiprocessing Linear Algebra Algorithms on the CRAY X-MP/2: Experiences with Small Granularity", Journal of Parallel and Distributed Computing, 1, 1984, pp. 22-31.

14. George, A., Heath, M. T. and Liu, J., "Parallel Cholesky Factorization on a Shared Memory Multiprocessor", Tech. Rept. ORNL-6124, Oak Ridge National Lab., March, 1985.

15. Farhat, C. and Wilson, E., " A Parallel Active Column Equation Solver", Computers and Structures Vol. 28. No. 2, 1988, pp. 289-304

16. Fulton, R. E., and Chiang, K. N., "Computational Crash Dynamics Methods for Fifth Generation Supercomputers". GM Project E-25-633 Final Reports Phase I, Dec. 1987.

17. Goehlich D., Chiang K. N. and Fulton R. E., Parallel Computer Approach to Finite Element Methods, UPCAEDM 1988 Conference, pp. 139-144, June 27-29, 1988, Atlanta, Georgia.

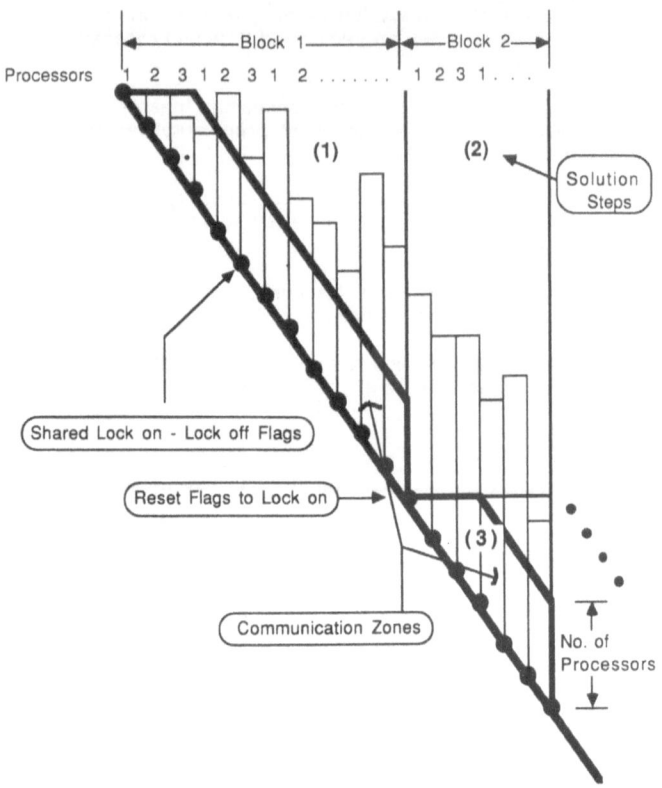

Figure 1. Skyline $LD^{-1}L^T$

	Maximum No. Processors	Vector	*Memory
High Speed Systems	** (2400 -		
Cray Y-MP	8 3600)	*	S
IBM 3090	6 (900)	*	S
ETA	10 (10,286) e	*	S/L
Massively Parallel			
BB&N GP1000	256 (125) **		L
INTEL iPSC/2 VX	64 (1028)	*	L
N Cube 10	1024(500)		L
Amatek 2010	512 (80)	*	L
Connection Machine	65,536 (2500)		L
Meiko Computing Surface	1000 and up (1000 or more)		Transputer
Intemediate Systems			
Alliant FX/80	8 (189) **	*	S
Elxsi 6420	10 (120)		S
Convex C240	4 (200)	*	S
FLEX/32	20 (80)	*	S/L
High Performance Workstations			***
Ardent TITAN	4 (64) **	*	S (128)
Stellar GS-10000	4 (40)	-	S (128)
Raster/Sun	8 (160)	*	L (128)
Apollo DN-10000	4 (144)	-	S (128)
Pixel	82 (820)	-	S (128)
Silicon Graphics	2 (40)	-	L (41)
S = Shared, L = Local	** MFLOPS	*** MBYTE Memory e - Estimated	

Table 1. Parallel Computer Family

Machine	Year	Number of Processors	Rate Speed MFLOPS	Memory MWORDS
CRAY-1	1976	1	160	1
X-MP/2	1982	2	420	4
X-MP/4	1984	4	840	8
CRAY-2	1985	4	1,700	128
Y-MP	1988	8	2,500	32
CRAY-3	1989	16	16,000	572
CRAY-4	1992	64	128,000	2048

Table 2. CRAY Supercomputer History/Plans

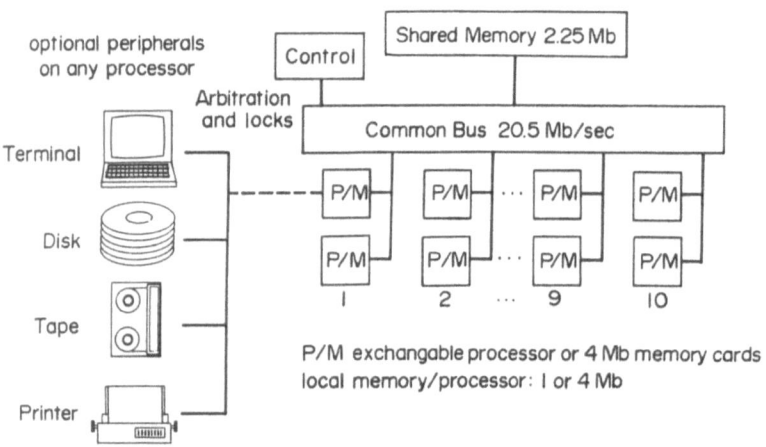

Figure 2. FLEX/32 Multicomputer

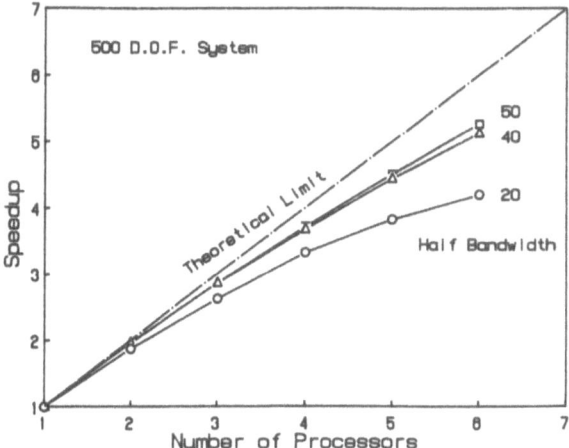

Figure 3. Cholesky Decomposition (FLEX/32)

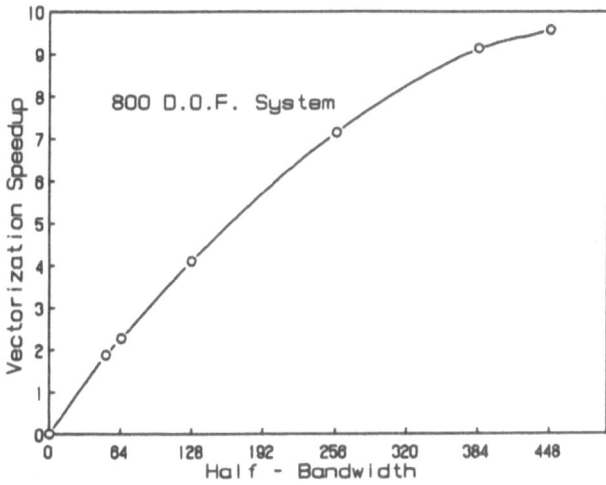

Figure 4. Cholesky Decomposition (CRAY X-MP/48)

Half-Bandwidth = 100, Total D.O.F = 1000

No. of Proc.	Wall Clock	Speedup
1	3.827 (sec)	1
2	0.89	4.3
3	0.686	5.58
4	0.568	6.74

Half-Bandwidth = 200, Total D.O.F = 1000

No. of Proc.	Wall Clock	Speedup
1	13.37 (sec)	1
2	3.01	4.442
3	1.67	8.006
4	1.25	10.7

Figure 5. Cholesky Decomposition Case I (IBM 3090-600E) Parallel/Vector Compare With Sequential/Scalar

Half-Bandwidth = 100, Total D.O.F = 1000

No. of Proc.	Wall Clock	Speedup
1	1.59 (sec)	1
2	0.89	1.79
3	0.686	2.32
4	0.568	2.8

Half-Bandwidth = 200, Total D.O.F = 1000

No. of Proc.	Wall Clock	Speedup
1	4.37 (sec)	1
2	3.01	1.46
3	1.67	2.62
4	1.25	3.5

Figure 6. Cholesky Decomposition Case II (IBM 3090-600E) Parallel/Vector Compare With Sequential/Vector

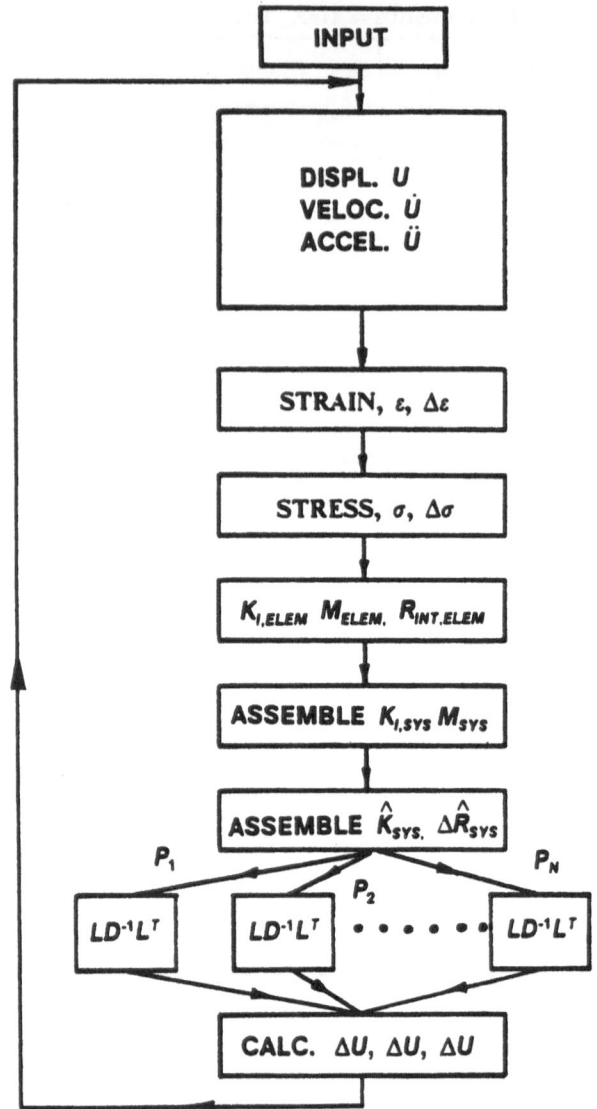

Figure 7. FENRIS/P1

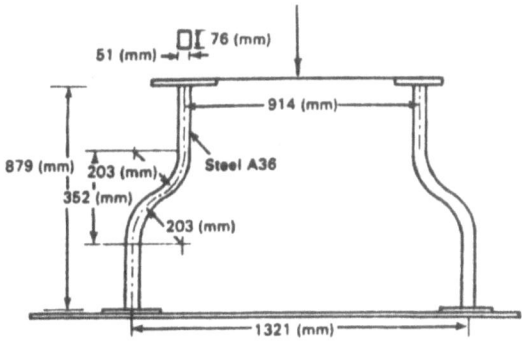

Figure 8. S-shape Torque Beam Subjected to Impact

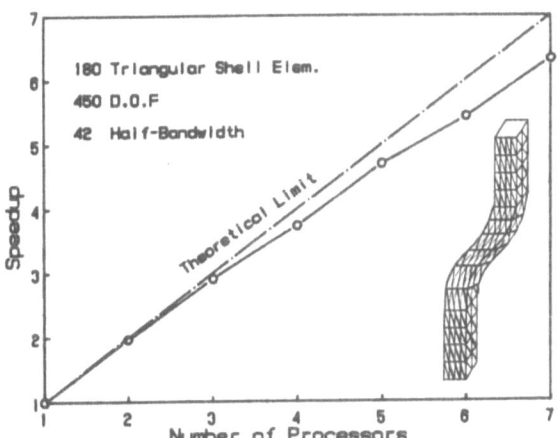

Figure 9. FENRIS Skyline Matrix Decomposition

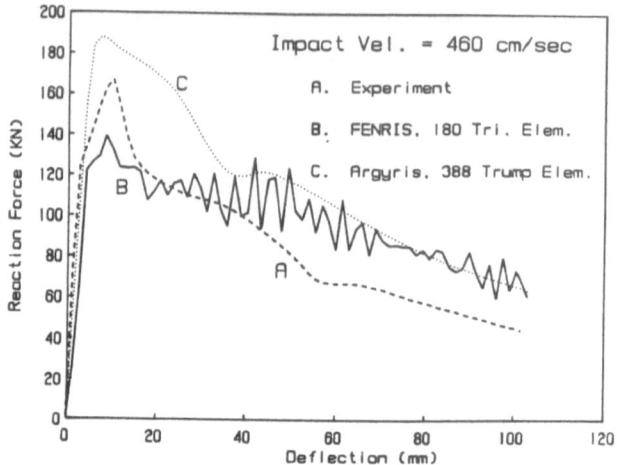

Figure 10. Force - Deflection During Impact

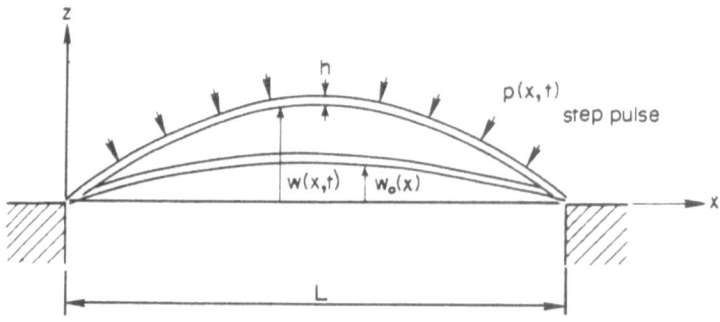

$$E I \left(\frac{\partial^4 w}{\partial x^4} - \frac{d^4 w_o}{dx^4} \right) - N \frac{\partial^2 w}{\partial x^2} + \rho A \frac{\partial^2 w}{\partial t^2} = - p (x,t)$$

$$N = \frac{EA}{2} \left[\left(\frac{\partial w}{\partial x} \right)^2 - \left(\frac{d w_o}{d x} \right)^2 \right]$$

Figure 11. Geometry of a Shallow Arch

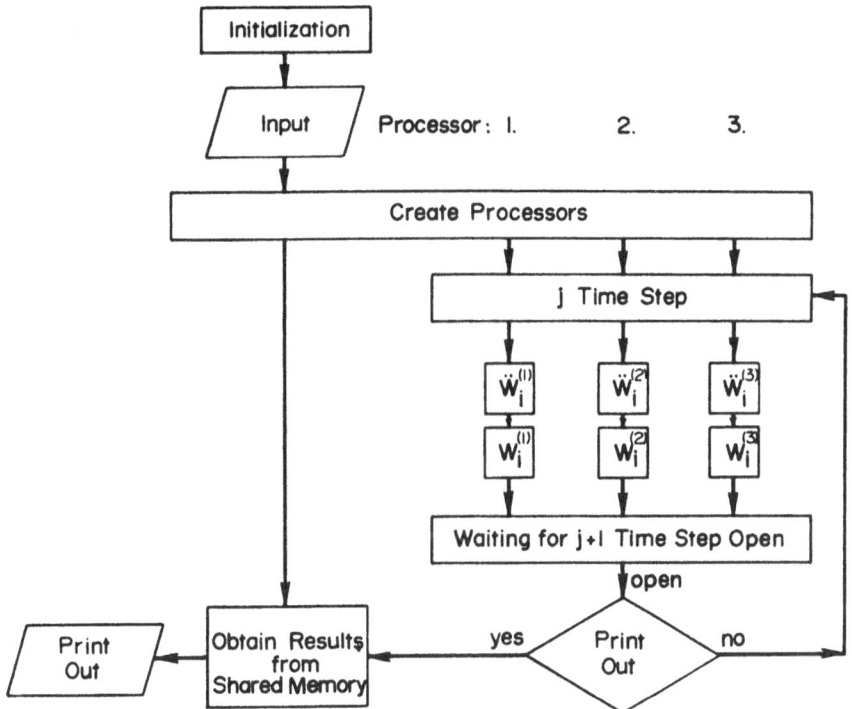

Figure 12. FLEX/32 Integration Approach

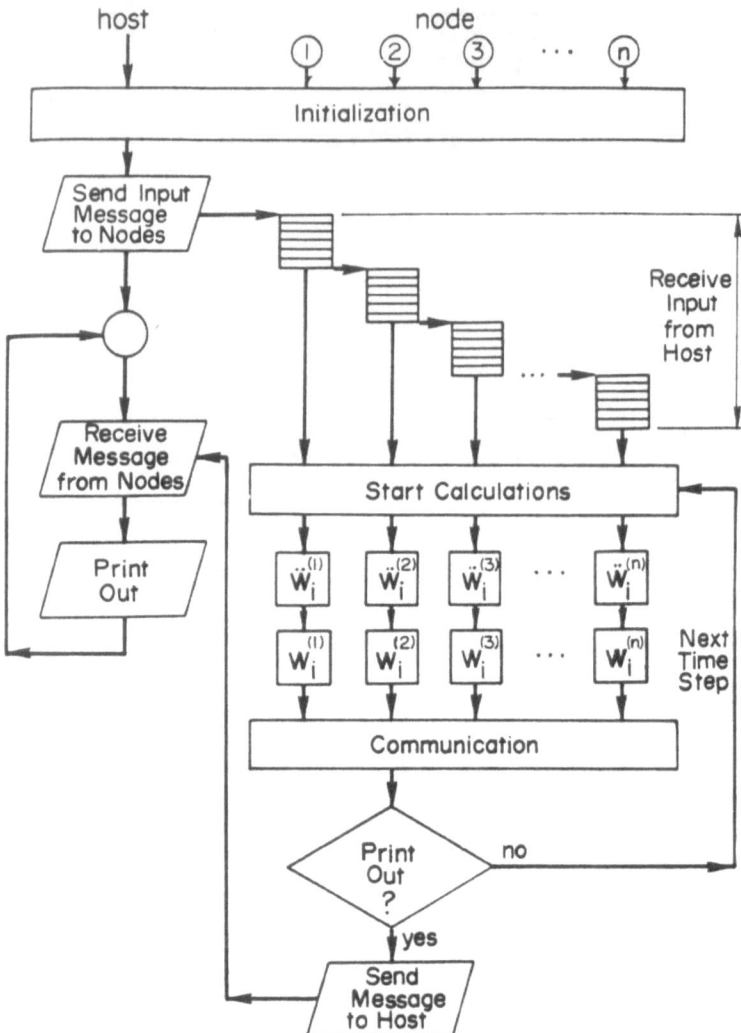

Figure 13. Intel iPSC Integration Approach

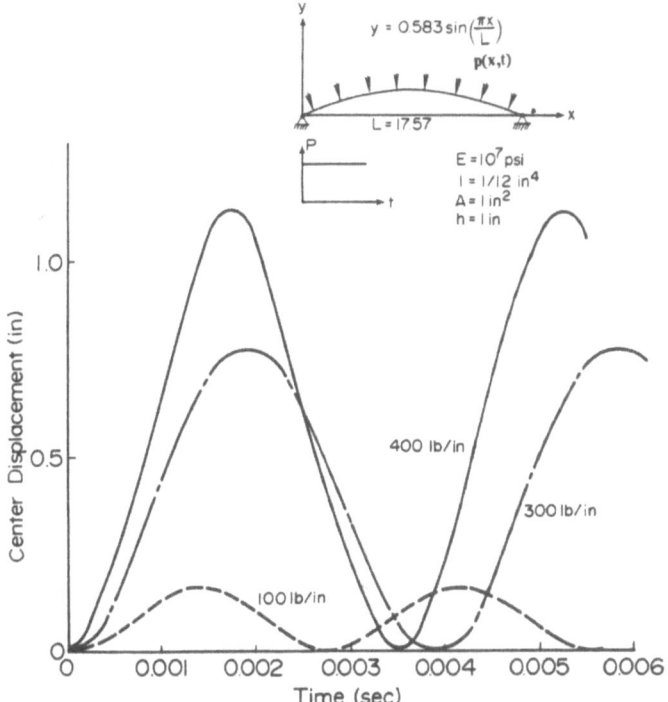

Figure 14. Dynamic Response of Shallow Arch

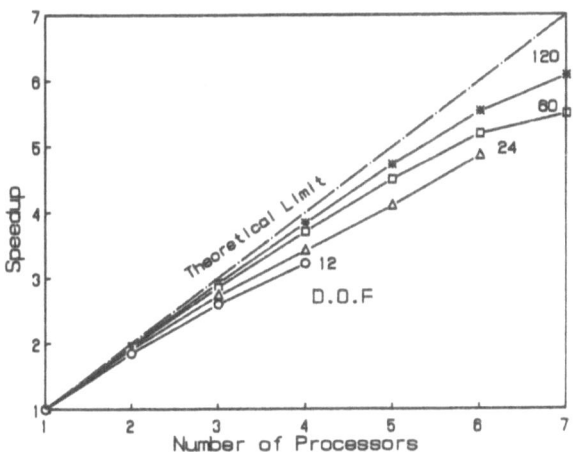

Figure 15 Speedup for Shallow Arch (FLEX/32)

Basic Performance of Two Graphics Supercomputers: Stellar GS1000 and Ardent Titan-2

K. Lue, K. Miyai

Institute for Supercomputing Research
8F Recruit Kachidoki Building
2-11 Kachidoki Chuo-ku
Tokyo 104 Japan

Abstract

The performance of the Ardent Titan-2 and the Stellar GS1000 computer are analyzed through two sets of standard benchmarks: the Los Alamos Vector Operations Kernels and the Lawrence Livermore Fortran Kernels. Both computers are designed to be graphics supercomputers, though, in this paper we are interested in examining them as computational engines. The overall performance of both systems as indicated by the LFK harmonic mean is quite similar, despite the large difference in architecture. Close scrutiny of the test results shows that these basic architectural differences allow them higher performance in certain benchmark kernels. Models are presented to explain these performance characteristics based on the machines' architecture.

Introduction

Since 1987 a new class of high performance single-user graphics computer has been gaining the attention of scientific research community. These machines typically come with the computing power of a minisupercomputer capable of a few tens of MFLOPS. What distinguishes these machines from other minisupercomputers are their integrated high performance graphics capabilities. Most of them are UNIX-based and allow both multiple and single user configurations. Their computational performance makes them affordable and extremely attractive to people with medium size computational problems because their performance to cost ratio is relatively high. As graphics rendering units, these machines are designed with large data paths to ferry data between the computational and rendering units. Special hardware is built in to perform fast graphics related transformations required in high speed visualization.

64

In this paper, we will look at the two computers as numerical problem solvers. Their respective architectures and how they affect the performance of the benchmarks ran on them are discussed. We will report the results obtained from the twenty-four Lawrence Livermore Fortran Kernels (LFK) and the Los Alamos Vector Operation Kernels (Vecops). These benchmarks are sufficiently detailed to give us a rough indication of the machines' performance because they range from basic vector operations to fairly complicated application cores.

Machine Architecture

The GS1000 is a shared memory machine with a main memory of 16 - 128 MB in increments of 32 MB. The computational engine consists of a multi-stream processor (MSP) and a vector/floating point processor (VFP). The MSP consists of four independent computational streams each capable of executing a common set of machine instructions either independently or in concert with each other. The machine clock is rated at 50 ns and every clock cycle one instruction can be issued to one of the four streams giving a collective performance of 20 MIPS. This translates to an effective clock of 200 ns for each stream.

The vector/floating point processor is shared by all four streams and runs at a clock rate of 50 ns. The unit consists of 6 vector registers each capable of holding 32 64-bit elements. At one floating point calculation per clock cycle, the vector/floating point unit is capable of performing 20 MFLOPS. On certain calculations, the vector unit can chain add-multiply floating point operations, giving a peak performance of 40 MFLOPS (*manufacturer's note: 40 MFLOPS is a hardware theoretical peak. In actual computation, a sustained peak is around 32 ~ 34 MFLOPS*).

The main memory is serviced by a data path and a 1 MB with a maximum data transfer rate of 1.28 GB per second. However, the memory bandwidth is only 320 MB/sec but is still sufficient to drive the VFP at peak performance. In a vector operation, a vector load instruction can fetch up to 32 double-precision numbers from memory which can be loaded into the vector register at a rate of 64 bytes every 200 ns. This translates to two double precision numbers per 50 ns cycle sufficient for the peak performance of 20 MFLOPS which requires two operands per cycle.

The GS1000 multi-stream processor is a single physical unit consisting of four logical processors known as streams. Since only one instruction is issued every 200 ns for any given stream, only one streams is active at any given clock cycle . The vector/floating point unit is shared by these four streams and when one of the streams initiates a vector calculation, the other three are placed in a wait state until the vector operation is completed. The vector unit, therefore, cannot be time-sliced. This allows the GS1000 to achieve a peak of 20 MFLOPS in vector operations even when only one stream is active.

The Titan is also a shared memory computer which uses a multiprocessor instead of multi-stream architecture. In its multi-processor configurations, the Titan can have up to four physically separate processors. Each processor runs at a clock rate of 125 ns. It is a register-to-register machine with 16 way interleaved memory serviced by a 256 MB/s bus. Each processor unit contains one integer processor and one vector processor (a Weitek chip, consisting of a floating point add, multiply, and logic unit). Each vector processor has a peak performance of 8 MFLOPS for single vector operations and 16 MFLOPS for combined vector multiply-add operations. The integer unit uses a 32-bit MIPS chip with a peak performance of 16 MIPS. Each processor contains 32 vector registers, each capable of storing 32 64 bit elements. The Titan's architecture is quite similar to that of the Cray X/MP's.

Benchmark Codes

One of the goals of computer benchmarking is to verify and examine the performance of a computer. Good benchmarks should allow one to reveal the strengths and weaknesses of a computer, from which one may extrapolate the likely performance of the machine under a different type of workload. No one benchmark can fulfill everyone's requirements and because of the wide range of possible applications, a truly universal benchmark is not possible. In this study, we are primarily interested in assessing the fundamental computational functions which involves their ability to do arithmetic operations, strided data operations, multiple array calculations etc. We would like also to relate the machine architecture to the observed performance. The benchmarks we have selected are in our opinion well suited to this purpose.

The Vecops kernels are a set of Fortran loops that perform basic arithmetic operations such as scalar-vector adds and vector-vector multiplies. Performance in these loops allows the user to examine the raw computational capabilities of the computer because of the simplicity of these loops. Under these test conditions machines are expected to operate close to their theoretical peak. To examine vector startup costs, the Vecops kernels also contain test operations with variable vector lengths. The Vecskip kernels are essentially the same as the Vecops kernels except that they are run with vector strides of 1, 2, 4, 8, 23, and 49. Results for stride one are the same as that of Vecops. These different strides help to identify the machine's performance degradation, if any, when non-unit stride memory accesses are used. This is an important aspect as most physical problems frequently require non-unit stride type memory accesses.

The Lawrence Livermore Kernels (LFK) are a little more complicated than the Vecops/Vecskip kernels. They are a set of twenty-four Fortran loops extracted from real applications used at Lawrence Livermore National Laboratory and were consolidated as representative of the workload at Livermore. Since they are only the computational cores of actual application codes, extrapolation of these results to predict the performance of any real applications should be done with care.

Benchmark Environment

We performed our tests on a two processor Titan with 64 MB of main memory with System V Unix version 1.1 and Fortran compiler version 1.1. The GS1000 tested also had 32 MB of main memory and ran Stellix V1.5.1 (the Stellar version of the Unix operating system) with version 4.0 release 11.0 of the Fortran compiler. The tests were performed in February 1989.

Benchmark Results and Analysis

Los Alamos Vector Operations Loops

To understand the basic capabilities of a computational unit, it is best to start with an analysis of its performance on very simple arithmetic operations. The Vecops kernels are a set of simple scalar-vector operations that involve only adds and multiplies. Results from these 13 kernels are listed in Table 1. First nine kernels have direct memory accessing while the last four use indirect memory addressing.

The first four kernels of the group contains single operations while the last five in the group examine multiple arithmetic operations. We ran these loops with various vector lengths of 25 to 1000 and the results reported here reflect that of vector length of 1000. Performance improves steadily with increasing vector length which asymptotes at vector length of 1000.

In scalar mode, the Titan performs at approximately 1.2 MFLOPS for the single operations (with unit stride data) while the GS1000 clocks around 0.48 MFLOPS for the same four kernels. The lower performance of the GS1000 can be explained by the way scalar instructions are issued. Though the GS1000 has a 50 ns clock, with its four

Machine	Titan				Stellar			
Option -O1	Stride 1	Stride 2	Stride 4	Stride 8	Stride 1	Stride 2	Stride 4	Stride 8
V=V+S	1.35	1.35	1.35	1.37	0.49	0.35	0.35	0.33
V=S*V	1.35	1.35	1.35	1.37	0.48	0.35	0.35	0.33
V=V+V	1.27	1.25	1.25	1.26	0.48	0.32	0.32	0.28
V=V*V	1.26	1.25	1.25	1.26	0.49	0.44	0.44	0.37
V=V+S*V	1.75	1.74	1.74	1.76	0.88	0.60	0.60	0.54
V=V*V+S	1.75	1.74	1.74	1.76	0.75	0.54	0.54	0.48
V=V*V+V	1.26	1.50	1.50	1.50	0.88	0.48	0.48	0.42
V=S*V+S*V	2.63	2.55	2.63	2.63	0.91	0.86	0.86	0.77
V=V*V+V*V	1.96	2.11	2.14	2.14	1.04	0.63	0.62	0.55

Option -O2	Stride 1	Stride 2	Stride 4	Stride 8	Stride 1	Stride 2	Stride 4	Stride 8
V=V+S	6.84	6.61	6.61	4.92	4.76	1.28	1.19	1.19
V=S*V	7.18	6.52	6.52	4.92	5.26	1.39	1.22	1.22
V=V+V	5.48	5.50	5.79	4.91	4.26	1.19	0.96	0.89
V=V*V	5.47	5.53	5.79	4.91	4.44	1.04	0.94	0.88
V=V+S*V	11.22	11.06	11.71	9.80	7.41	2.04	1.89	1.70
V=V*V+S	10.98	11.45	11.29	9.80	7.69	2.00	1.85	1.67
V=V*V+V	6.69	6.43	6.60	5.07	6.45	1.67	1.56	1.37
V=S*V+S*V	10.81	10.52	10.58	7.62	9.68	2.94	2.73	2.50
V=V*V+V*V	7.03	7.09	6.17	5.89	8.00	2.03	1.92	1.47

Table 1.

stream architecture each stream sees only one-every four cycles. This means that when running in scalar mode (single stream), the GS1000 has an effective clock of 200 ns resulting in lower scalar performance.

In vector mode, Running the kernels in vector mode with contiguous data the Titan ran at 5.4 and 7.1 MFLOPS. The GS1000 on the same single kernels ran between 4.2 and 5.2 MFLOPS On the more complex kernels with more than one vector operations the Titan peaks at 11.2 MFLOPS for stride one data while the GS1000 on similar kernels has a performance maxima of 9.6 MFLOPS. Below we will present models for the Titan and the GS1000 which attempt to explain the consistency of this data.

With clock speeds of 50 ns and 125 ns (for the GS1000 and Titan respectively), the theoretical peaks of these machines should be 20 and 8 MFLOPS for single vector operations of the form $V = V + V$. When combined operations multiply/add are possible the theoretical peak should be doubled at 40 MFLOPS and 16 MFLOPS. The results reported above are therefore much lower than the expected peak values. For vector-scalar add and multiply operations, the Titan's actual performance is above 80% of peak performance and for vector-vector multiply and add it runs at 67% of its peak speed.

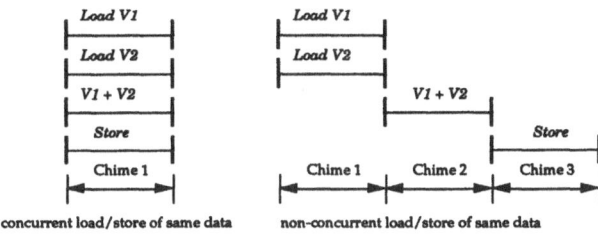

FIGURE 1. Vector operation $V = V1 + V2$.

For single vector operations, the GS1000 seems to perform at a quarter of its peak rate, which is too low for a machine working on such simple operations. Our results suggest a bottleneck exists which limits the GS1000's performance. For a rate of 20 MFLOPS, we must assume that the GS1000's supporting architecture is capable of channelling two operands to the vector floating point unit and storing one result out to the vector registers, all within a single chime. We must also assume that the two floating point chips (each with a 100 ns clock) can be perfectly synchronized to run concurrently giving an aggregate performance of two results every 100 ns (i.e. one result every 50 ns). The GS1000's vector load and store paths have sufficient bandwidth to perform the data movement, however, if these loads and stores cannot be completed within the same chime as the vector operation, the overall performance will be adversely affected. In Figure 1, we present a model in the form of a chime sequence for a single vector operation. When a concurrent load/store are chained together with the vector operation, the GS1000 can achieve its peak performance of 20 MFLOPS for single vector operation. On the other hand, the inability to chain load and store operations forces the system to use two extra chimes for loading and storing. This cuts performance by a factor of three, resulting an effective peak performance of 6.7 MFLOPS. The bottleneck does not exist within the VFP unit itself: it is the inability to chain the load-compute-store operation that is the factor reducing the performance.

This analysis can be extended to the more complicated kernels in the set. Considering the last kernel, $V=V^*V+V^*V$, our model shows that the peak performance in this case is 15 MFLOPS (Figure 2.). During the first chime, the first two vectors can be loaded and since multiplication cannot take place the unit must wait until the next chime. While performing the first multiplication, the system loads the next two vectors. In the third chime, the full combined vector multiply and add capability of the vector unit is fully realized. In this chime, the addition can proceed as soon as

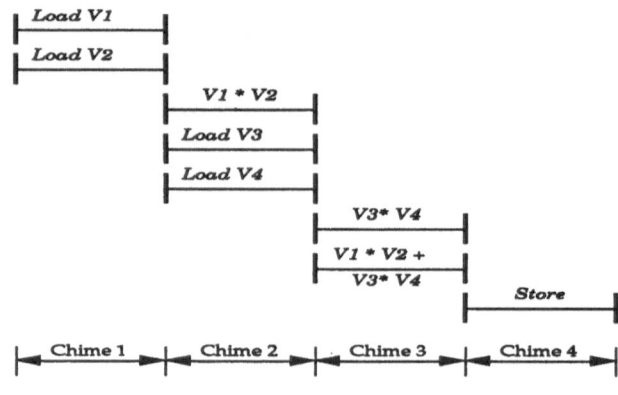

3 operation/4 Chimes = 15 MFLOPS peak

FIGURE 2. Vector operation $V = (V1 * V2) + (V3 * V4)$.

results are computed. As the add and store operation cannot be executed in the same chime for the same data, a fourth chime is then required for the final store. This vector operation therefore needs four chimes for three arithmetic operations giving a theoretical limit of 15 MFLOPS (the measured performance of 8 MFLOPS for this kernel is therefore 53% of the peak performance).

The observed performance on the Titan can be explained by a different kind of bottleneck. The data transfer rate in and out of the vector registers is 512 MB/second, which translate to 8 words every clock cycle. With the two vector/single scalar operation, $V=V*V+S$, the effect of this bottleneck is not obvious. For this operation, we have recorded 10.9 MFLOPS which is approximately 68% of the peak 16 MFLOPS performance. The vector register bandwidth limitation is more clearly felt in the last kernel, $V=V*V+V*V$. To understand the effect of the bottleneck in this case, we present a slightly different chime diagram. Figure 3 shows a 'snapshot' within each chime, showing what is happening during the computation cycle.

In the first chime, the arrays V_1 and V_2 are loaded from memory to the vector registers. This takes up two of the 8 word/chime bandwidth. Channelling these operands from the vector registers, performing the add, and storing the result back to a third vector register take up another 3 words. At this point V_3 and V_4 cannot be loaded for further operation because only two load pipes are available. The next operation can only be initiated in a second chime. In the second chime, V_3 and V_4 are loaded and multiplied. With the chaining capability, the temporary result from the previous operation is fetched and used in the addition. At this stage, there is only

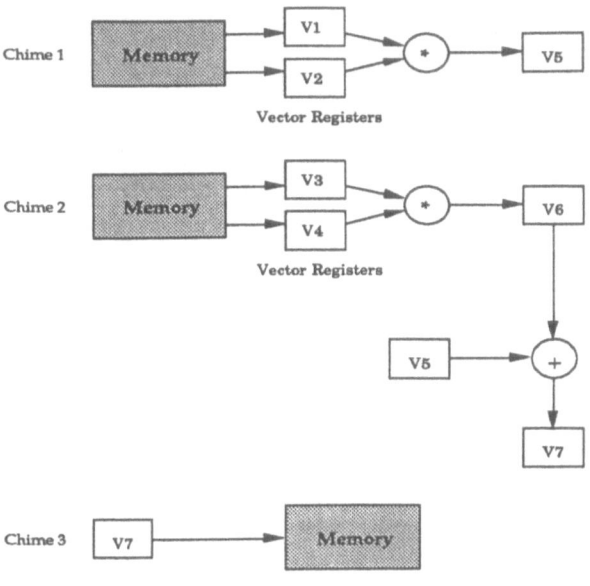

3 operations / 3 Chimes = 8 MFLOPS

FIGURE 3. Vector operation $V = (V1 * V2) + (V3 * V4)$.

enough bandwidth to move the result from the adder to a vector register. A further store back to memory cannot be performed as the 8 word bandwidth of the vector registers has been completely utilized. The last vector register to memory store must be initiated in completely new chime. In an actual compiler optimized code, the load within each chime might be more evenly divided, e.g. the addition may be moved to the final chime just before the store operation. Therefore, for this kernel, three chimes are required to perform the 3 arithmetic operations instead of the expected two. This gives a peak performance of 8 MFLOPS. Our measured 7.03 MFLOPS performance on this kernel is then quite reasonable.

As pointed out earlier, the Vecskip kernels are computationally the same as that of the Vecops kernels, except that they access memory in strides of 2, 4, and 8. The fact that most physical problems often lead to non-unit stride accesses motivated us to investigate this area. Results for these tests are also presented in Table 1.

For scalar operations (option O1), Titan's performance virtually remains the same across all tested kernels with the exception of the $V=V+V*S$ which showed a 20 percent increase over unit stride in performance with stride 2, 4, and 8 accesses. On the other hand, strided data seemed to have a larger effect on the GS1000. In general, strided data results in a performance decrease of more than 20 percent for the GS1000 in scalar mode.

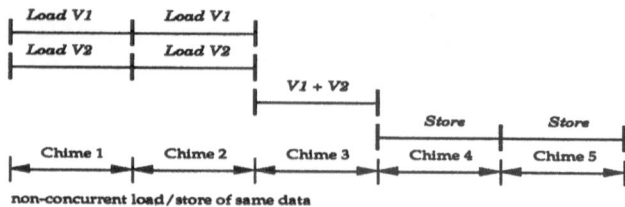

FIGURE 4. Vector operation $V = V1 + V2$ (Stride 2).

The Titan's vector performance on Vecskip does not vary much compared to its unit stride performance. However, a dip in performance is noticeable for stride 8 accesses, due to the way the memory banks are interleaved. The Titan memory bank uses 16 way interleaving. Memory conflicts occur whenever we have stride eight accesses. The GS1000 is 70 ~ 80% slower on non-unit strides, and the decrease observed is quite uniform across all kernels. The degradation can be traced to the use of the cache in data transfer. Though the GS1000 has a large data bandwidth, non-unit stride accesses forces it to fetch data that the VFP does not need during non-unit stride operations. In the stride 2 case, every other operand it fetches is unnecessary. Our model in Figure 1 can be extended to take this into account. In essence, stride 2 data accesses dictate that the vector load needs to be twice as long for the same amount of computational work. The additional cost of masking during loading and storing of the operands and results also imposes some overhead. Without knowing the penalty of the masking operations, we can only predict a high end asymptotic peak performance of the GS1000 in the non-unit stride case. Figure 4 illustrates our model.

The vector load would take twice as long because of the stride and similarly for the store. We have ignored the time required for masking in this model. With this, the operation $V=V+V$ takes five chimes to complete, giving an effective theoretical peak of 4 MFLOPS. If masking is taken into account, this theoretical peak would be lower. The performance for stride 2, in this case, is approximately 25% of our predicted peak. It is therefore highly possible that masking is required.

Lawrence Livermore Fortran Kernels

Proceeding on to look at more complicated kernels, we next examine the Lawrence Livermore Fortran Kernels (LFK). Table 2 is a summary of the scalar and vector

	TITAN	**GS-1000**
scalar		
Maximum rate	2.343	2.274
Average rate	1.243	0.957
Geometric rate	1.025	0.862
Harmonic mean	0.811	0.779
Minimum rate	0.213	0.380
vector		
Maximum rate	11.724	21.026
Average rate	3.633	3.947
Geometric rate	2.174	2.150
Harmonic mean	1.253	1.312
Minimum rate	0.234	0.370

Table 2.

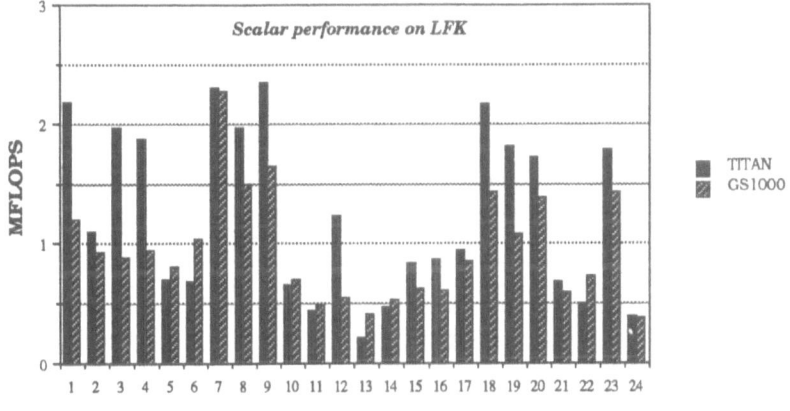

LFK Kernel Number

FIGURE 5. Scalar performance on LFK.

results (average vector length=471). The harmonic means of both machines are very similar, with the Titan at 0.811 MFLOPS and the GS1000 at 0.779 MFLOPS (in scalar mode). In vector mode, the GS1000 is slightly better (1.312 MFLOPS) than the Titan (1.253 MFLOPS). The slower performance of the GS1000 in scalar mode is due to a slower 'effective' clock of 200 ns compared to Titan's 125 ns clock, while its better

74

performance in vector mode is attributed to its excellent performance on kernel 22 which will be discussed later in this section. When compared to the Cray X/MP these harmonic means indicated that both systems are performing at approximately one eleventh of the Cray speed in scalar and vector modes.

In general, the various types of means reported in Table 2 did reflect performance similarities in both machines. However, single number statistics often smooth out the distinctive system features or weakness which we are trying to identify. In the next few paragraphs, we would review some of the more important kernels in the set in an attempt to explain how the GS1000 and Titan function. We will also apply the models discussed in the Vecops section to explain the performance observed.

Figure 5 shows the scalar performance of the GS1000 and Titan on all twenty-four kernels. In this mode, the Titan is faster in kernel 1, 3, 4, 12, 18, and 19. This result can be traced to the fact that these kernels contain predominantly operations of the form $V=V+V$ and $V=S+V^*V$ which we found to run faster on the Titan in the Vecops tests. In particular, kernel 3 is an inner product kernel of the form $S=S+V^*V$ and

LFK Kernel	Titan	GS1000	VP400
#1	V	V	p
#2		**	
#3	V	V	V
#4	V	V	p
#5		**	
#6	V	**	V
#7	V	V	V
#8	V	V	V
#9	V	V	V
#10	V	V	V
#11		**	p*
#12	V	V	V
#13	p	p	p
#14	p	p	p
#15		V	p
#16			
#17			
#18	V	V	V
#19		**	
#20			
#21	V	V	V
#22	V	V	V
#23		p	
#24			

V   vectorized
p   partially vectorized
**   not enough vector code
p**   partial vectorization was possible, but the over head is too large.

Table 3.

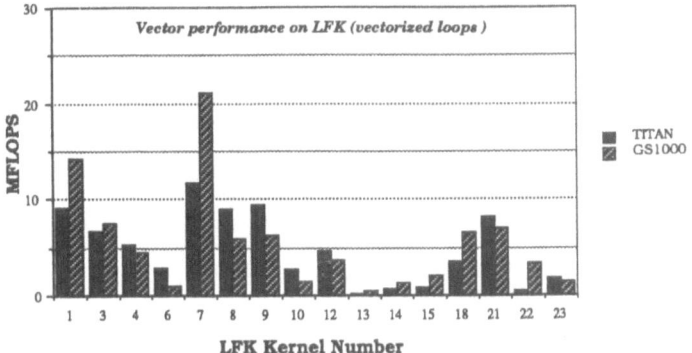

FIGURE 6. Vector performance on LFK (vectorized loops).

kernel 4 contains strided accesses which degrade the performance of the GS1000. In vector mode operations, the compiler plays a crucial role in the final performance. Table 3 displays the autovectorization results from the two compilers. We notice that kernel 6 can only be vectorized on the Titan though it fails to vectorize kernel 16. The inability of the GS1000 to vectorize kernel 6 was due to the variable vector length of the inner most loop: the initially short vector length of this loop prevented vectorization.

Performance of the kernels vectorized by either system are presented in Figure 6. In vector mode, the GS1000 is faster on kernel 1, 7, 14, 18, and 22. The Titan, however, is faster on kernel 6, 8, 9, and 10. For these last three kernels, the lower performance of the GS1000 is a result of non-unit stride accesses, explained in earlier sections. In the seven kernels contain strided memory accesses (kernels 4, 8, 9, 10, 11, 13, and 21), the Titan's performance is better except on kernel 13, which contains a MOD function. In this kernel, the Titan's performance edge is lost due to the slow execution speed of the MOD function.

We are particularly interested in the result of kernel 22 where the GS1000 is approximately six times faster than the Titan. This kernel contains division and exponent calculations which execute much faster on the GS1000 than on the Titan. In Figure 7. and 8., we present the performance of the two machines on division and exponential operation. *(A new Beta version (V2.0) of the Titan compiler was tested and the result from kernel 22 on this new compiler was timed at 1.71 MFLOPS)*

Another point of interest from Figure 6 is the excellent performance of GS1000 on kernels 1 and 7. Recall that the GS1000 is slower than the Titan on simple vector operations, which is explained by its inability to do chained load, compute, and store operations on the same data. However, for more complicated kernels like #1 and #7,

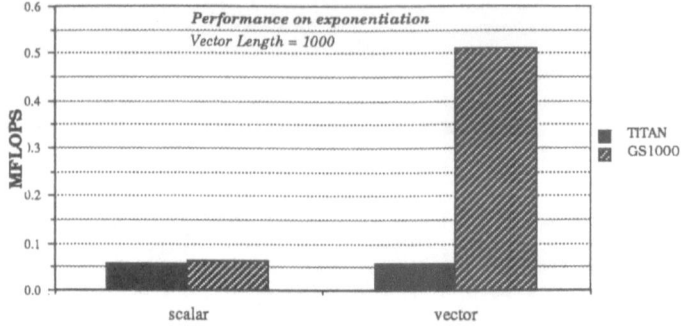

FIGURE 7. Performance on exponentiation (vector length = 1,000).

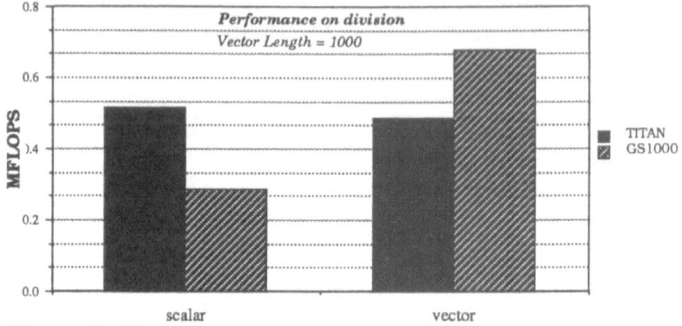

FIGURE 8. Performance on division (vector length = 1,000).

the GS1000 is able to compensate for this deficiency by processing different data at different stages of the load-operate-store phase. This is illustrated by extending our previous analysis on Vecops to kernel 1 in Figure 9 and 10.

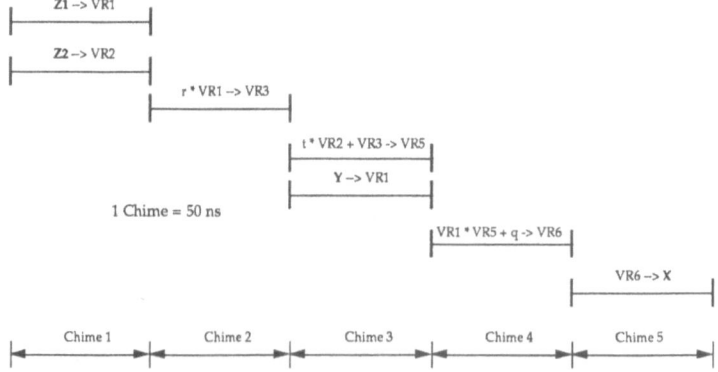

FIGURE 9. Chime sequence for Kernel 1 (GS1000).

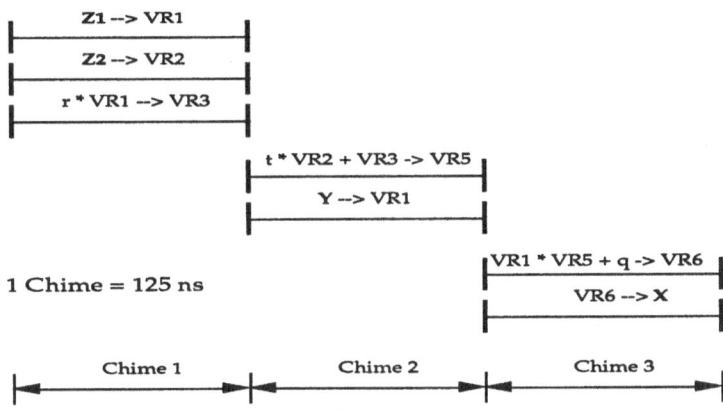

FIGURE 10. Chime sequence for Kernel 1 (Titan).

Kernel 1 consists of the following operations: $X = q + Y * (r * Z1 + t * Z2)$. (bold face characters indicate vector quantities). The GS1000 requires five chimes to complete the operation while the Titan requires only three. Thus, we can predict a theoretical upper performance bound for both machines,

5 operations / (5 chimes x 50 ns) = 20 MFLOPS peak (Stellar)
measured performance = 14.26 MFLOPS.

5 operations / (3 chimes x 125 ns) = 13.33 MFLOPS (Titan)
measured performance = 9.17 MFLOPS.

By carrying through the same analysis on kernel 7, $(X = U1 + r * (Z + r * Y) + t * (U2 + r *(U3 + r* U4) + t * (U5 + r * (U6 + r * U7)))$, we find that for 16 operations, 11 chimes are needed on the GS1000 while the Titan requires 9 chimes to complete the same operations. The theoretical peak rates are ,

GS-1000: 29.09 MFLOPS peak (21.06 MFLOPS measured)
Titan: 14.22 MFLOPS peak (11.72 MFLOPS measured)

Compensation of its inability to chain load/store operations is not the sole reason to GS1000's speed on kernels 1 and 7. The fact that the VFP unit has a 50 ns clock contributed significantly to its performance. It is also entirely possible that by reducing memory accesses through clever use of the vector registers to store intermediate results would further enhance the computational speed.

Summary

We have presented benchmark results of the Ardent Titan and the Stellar GS1000. We have chosen two sets of benchmarks with increasing code complexity to look at how the performance characteristics of two machines. We found that, in most cases, the Titan performs better in terms of basic operations as indicated by the Vecops data. However, as the computation becomes more complicated, the Titan's computational advantage is offset by the GS1000 faster clock and its ability to make up for its deficiency in load/store chaining. For both computers, the speed of the vector floating point unit is not the only consideration which determine final performance. As we have shown, vector register bandwidth, chaining capability, data channel structure, and intrinsic libraries all contribute to overall performance. The models we have presented for both computers are by no means exact. They are introduced here merely to help us understand some of the major processes during a computation and to explain some results we observed. A potential user should always consult benchmark data with extreme care as one can only examine a small parameter space of the machine with any set of benchmarks. A true application is an integral of many areas from physics, algorithms, programming languages, machine architectures and even as far as programming styles. It is the intricate relations between all these complex components of computer hardware/software that make interpretation of benchmark results more of an art than an exact science.

Acknowledgement

The authors would like to thank Asahi Stellar of Tokyo for the use of their GS1000 and to Mr. Timothy Stewart of Stellar Computer Inc. for his useful discussions on the Stellar GS1000.

Bibliography

1)   E. Misaki, K. Lue, and R. Mendez, Preliminary Study of the CPU Performance of the Titan Compared with that of the Cray X-MP/1, Institute for Supercomputing Research, TR 88-11a

2)   Stellar Graphics Supercomputer System Overview

3)   Titan Technical Overview, Release 1.0

4)   Olaf M. Lubeck, Supercomputer Performance: The Theory, Practice, and Results, Los Alamos National Lab Report LA-11204-MS

5)   Frank H. McMahon, The Livermore Fortran Kernels: A Computer Test of the Numerical Performance Range, Lawrence Livermore National Lab Report UCRL-53745, December 1986

6)   John M. Levesque and Joel W. Williamson, A Guidebook to Fortran on Supercomputers, Academic Press, Inc., 1989

7)   Sidney Fernback (Edt), Supercomputers Class VI Systems, Hardware, and Software, North-Holland, 1986

High Bandwidth Interactivity and Super Networks

Mr. James "Newt" Perdue
Vice-President, Business Development

Ultra Network Technologies
101 Daggett Drive
San Jose, CA 95134

Abstract

A scientist's productivity is often limited by the tools at his disposal. Supercomputing has provided the researcher with a very valuable tool for solution of complex, multi-variable, multi-dimensional problems. However, the ability to become productive still depends on getting access to the results in a timely manner. Advances in network performance have lagged considerably behind advances in Supercomputing performance. In fact, over the past 10 years, supercomputing power has increased greater than a factor of 10 with no real increase in network performance. The advent of high resolution graphics technology has created a new toolbox for the visualization of results. These graphics tools require high performance networking in the gigabit per second range to permit practical applications to be developed. The use of high performance networks to achieve *high bandwidth interactivity* is explored here in various applications. The factors that affect network performance in the *gigabit/sec* range are described. Further, the architecture required to support such applications is presented in the context of today's network standards and technology, including ISO TP4, TCP/IP and FDDI.

The Growing Need for a Super Network

The explosive growth in the power and capacity of scientific computer systems has caused a real need for higher performance networks. For instance, over the last 10 years, the computing power of supercomputers has increased by a factor of 10 and the memory capacity has increased by a factor of 256 (Cray 1 in 1977 to Cray 2 in 1987). The capacity to generate data that must be sent over a network has increased dramatically while in that same timeframe, the performance of vendor provided networks has shown almost no growth.

In this same timeframe, pressures on the use of networks has increased substantially. Distributed processing is now accepted as the norm in most scientific computer centers. The proliferation of affordable workstations and the revolutionary advances in graphics technologies which these workstations have fostered puts further pressures on the requirements for increased network performance.

Networks have become increasingly important in producing scientific results. The scientific user typically interfaces with the major computing capability using a workstation where he can control and customize his work environment to his particular application and habits. His large data sets are stored on file servers specialized in data storage and management, and his processing is done on a selection of processors depending upon the tasks at hand. He may use a specialized processor for model building, a supercomputer for simulation and yet another computer for post-processing and graphics presentation. Yet he sits at one terminal

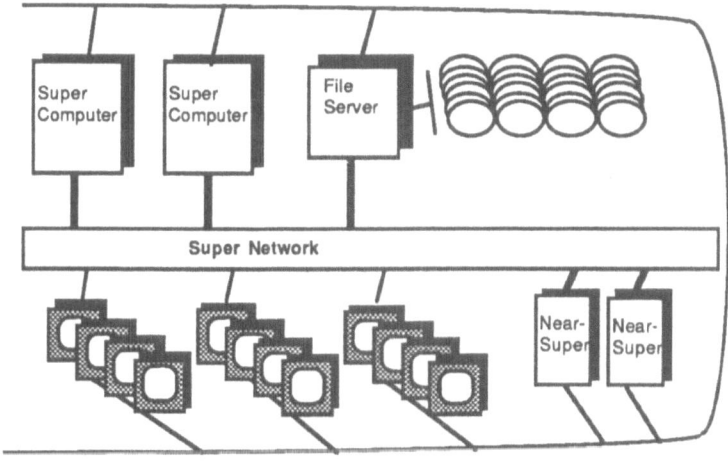

FIGURE 1.

directing the entire computational scenario. His ability to be *efficient* in using these systems is very often limited by the data network.

Traditionally, the problem-solution cycle for a scientific user of supercomputers has looked like this:

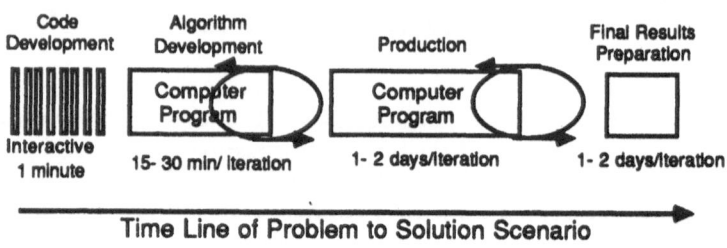

Time Line of Problem to Solution Scenario

FIGURE 2.

Typically, it has been very batch oriented, with some interactive processing during code development and job set-up. Programs take several minutes to hours to run and results are generally not queried until long after the computer has completed the run. In fact, many times the results must be run through output processors before the researcher can study them. This process takes literally months to get a scientific algorithm developed to the point that the researcher is dealing with new science rather than the logistics of computing. In this mode, the scientist's train of thought is interrupted many times for long periods while the computing scenario is performed, over and over again. In fact, we count hours and days between minor results. This must certainly interrupt and play havoc with innovative thought processes.

The "thought gaps" which this "batch oriented" scenario introduces is a real detriment to the efficiency of the researcher.

Closer interaction between the researcher and the computing tools should dramatically improve his efficiency. If he were to play "what if" games in a timeframe that fit his own thinking timelines he would likely make faster progress. Such a scenario is represented below.

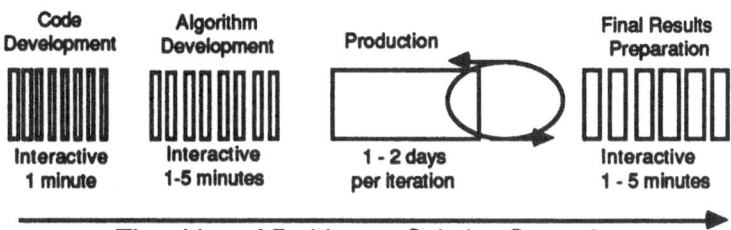

Time Line of Problem to Solution Scenario

FIGURE 3.

In this scenario, during algorithm development, the researcher is interacting with his program as it computes. He is able to try different parameters and their effect on the solution without waiting hours or days to examine the results. In fact, with the proper visual presentation of his results, he can assess the usefulness of any single iteration immediately. He can save time if he can see and assess the "early results" and stop the iterative process before it uses more of his (and the computers) valuable time. The simulation and analysis of his results can proceed simultaneously within this scenario.

In the process of final results preparation, he can be interactive in the development of movies, images and other presentation materials to disseminate his results.

By providing tools which improve his continuity of thought, he can follow intuitive thoughts, insights and creative impulses to their logical conclusion immediately.

What is required to move the researcher from the "batch oriented" to the "interactive oriented" development mode? The programs which he runs perform billions of computations prior to meaningful results. The results span potentially millions of computation points. How can he possibly interact with that much data in a short period of time?

The key to this lies in the following areas:

- faster supercomputers

- new approaches to algorithm development

- high resolution and very fast graphics devices

- high speed networks.

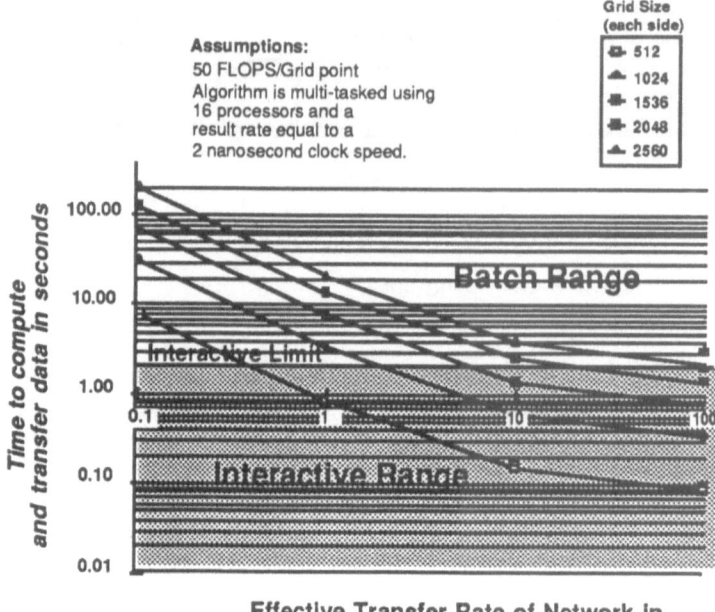

FIGURE 4. Network rates as a function of grid size for a 1990 supercomputer for algorithm animation or interactivity.

Today, supercomputers can perform work at the speed of one billion floating point operations per second. Within the next few years, with the continued growth in parallel architectures and tools, we can expect tens to hundreds of billions of operations per second available to the researcher. New algorithms will be necessary which output meaningful intermediate results for interactive use and take advantage of new methods to display them. The graphic display devices will have to be capable of displaying a complete field of information received from the supercomputer, render it into a realistic shaded model of a physical entity and display it in a second or less. And finally, networks will have to be capable of delivering the data to the graphics devices fast enough to permit display at rates anywhere from 1 per second to 30 images per second for animation.

Although today's computers permit interactivity, true high bandwidth interactivity must wait another year or two until the computers can compute and *display* both the physics and the image data in a period between one tenth second and one second for realistic grid sizes. The chart above assumes that a 1990's computer can compute at a rate of 8 Giga FLOPS and that a problem requires at least 50 operations per grid point. The chart describes the interactivity achieved as a function of the network rate for a set of grid sizes. For even a small grid size of 512 x 512 elements, a network rate of 100 MBytes/second is required to achieve 10 frames per second animation.

The chart below diagrams the required bandwidth for animation at just 15 frames per second as a function of both image size and color resolution. For today's monitors and true color (24 bits), it requires about 60 Million Bytes per second to get 15 frames per second. For just one image at 1280 x 1024 resolution, it requires 4 Million bytes. This greatly exceeds today's networks which typically get only 1 to 10 Million BITS per second throughput.

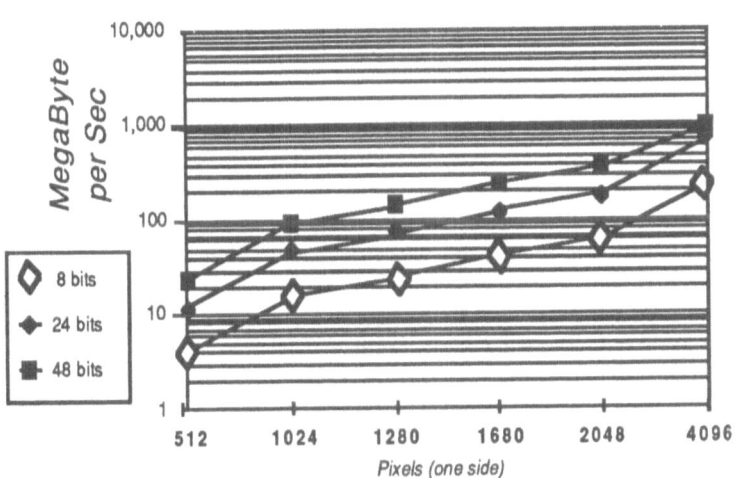

Transfer Rate Required For Animation of Image Data at 15 fps

(as a function of spatial and color resolution)

FIGURE 5.

Clearly, to support a high bandwidth interactive capability with future supercomputers, the capability to move data between the processors and user's graphics devices must improve dramatically.

Other Factors Driving Network Performance

Interactive processing is one major application of a high speed network. Technology and other applications continue to push for additional performance:

- Processor Speeds

The proliferation of new architectures and new technology permit results to be computed at rates exceeding 1 billion instructions per second today. It is possible to predict that this will increase by a factor of 10 to 100 over the next five years as we learn to combine parallel processing and new computing technologies.

- Memory Density and Speed

Memory chips contain over 1 million bits today. Within the next 2 - 3 years, 4 million bit chips will be commonplace. The Cray 2 is now available with 4 billion bytes of memory. Within 4 years, this may quadruple to 16 billion bytes. The size and speed of processor memory is important in determining the quality and resolution of the scientific results. Closer grid spacing on physical models improve the accuracy of simulation results from weather, aerospace, automobile to weapons research. This closer spacing requires more memory and access speed to match the faster processors. Programs will be capable of generating output that requires 20 Gigabytes of data per iteration.

- Higher Speed Data Sensors

Sensors in oil fields, space stations, auto tests, wind tunnels, medical instruments, and other applications generate billions of bits of information per second.

- Workstation Capabilities

Workstations provide the power our supercomputers did only 10 years ago. Distributed computing strategies which integrate the workstation processing capability with that of the supercomputer will demand faster communication between the two. Workstation memories that contain 16-32 million bytes of data will require networks capable of swapping the entire memory between computers in times close to 1 second. With tens to hundreds of these on a network, billions of bytes must be moved every second to maintain an adequate response time.

- Graphics Display Capabilities

Research in graphics has increased the ability to generate realistic graphics (shaded, rendered) from results output by the supercomputer. Powerful graphic processing engines now exist and more powerful ones are on the drawing board. The ability to transform millions of polygons per second will require the network throughput in the millions of bytes per second to support interactive processing. Graphic images in digital form will be the natural method to share information between researchers, as well as, the method to interact with the simulations.

- **Storage Facilities**

As new products for storage of information improve (in capacity, cost and performance), more data will reside on-line and the demand will increase for data retrieval due to the ability to combine this information into meaningful forms. New data search and retrieval algorithms demand faster access to more data from all users on the network. Information storage servers that require several Trillion Bytes of data on-line will be common in the next few years. The move toward disk stripping and parallel access drives now provides tens of megabytes/sec sources for data retrival and distribution.

- Communications Links

As fiber optic links become more available (lower cost and higher performance), remote users will demand the same quality of service as the local users. Fiber links that can transmit data at rates up to 4 gigabits/second will soon be commercially available.

- Application Algorithms

New science algorithms are being developed to take advantage of the new technology. High Bandwidth interactive computing will permit researchers to do "what ifs" with complicated algorithms and use sophisticated graphics systems to "visualize" the results.

Applications Requiring Super Networks

Clearly, with data being generated by super processors in chunks of billions of bytes, networks with effective throughput in the order of one million bits per second are NOT adequate. Interactive data requirements which require megabytes instead of a few characters and animations requiring gigabytes for animation prove this further. Small, incremental increases in network transfer capabilities won't be adequate.

The solution to the network problem will be closer when we can measure EFFECTIVE performance in the range of ONE GIGABIT PER SECOND.

Today's network products clearly do not provide the necessary bandwidth and throughput for this emerging set of requirements. New scientific applications demand that the performance of networks increase by factors of 10 to 100 almost immediately. If this is not provided, large imbalances in the systems used for research will occur. Scientific progress will definitely be slowed beyond acceptable limits. A few examples of such applications include the following:

- in *numerical fluid dynamics,* high resolution graphics will be the only way to explore the many alternatives to configurations for the new space plane project. For example, the ability to animate the fluid flow around the plane, change parameters and refine the model within minutes will be required to meet cost and schedule objectives for this project.

- in *medicine,* the ability to find new drugs for treatment of cancer, AIDS, and other highly dangerous diseases rest in the power of the computer to simulate and interact with the medical scientist. Formulation of new protein structures and configurations to attack viruses can be greatly improved with the use of powerful computers and high resolution interactive graphics. Further, new diagnostic tools, like the CAT and PET scanners need supercomputer power and super network performance to adequately reduce the billions of bytes of data in close to real time.

- in the *oil industry,* the ability to visualize reservoir dynamics with the aid of a skilled analyst may be the key to finding new sources of energy. Again, the network must supply the enormous amounts of input data and support the interactive visualization of the results.

- new *space systems* require a tremendous amount of simulation and prototype activity to ensure a robust system. High speed networking will pass the simulated sensor data between elements of a large heterogeneous computer environment. Research in support of the Strategic Defense Initiative and the Space Station require such simulation and networking.

- *many branches of science* require rapid evaluation and viewing of millions of images. A prime example is the billion of bytes per day sent from space sensors. These images will be enhanced on one computer, stored on another and displayed on yet another. Instant response to megabyte images is necessary to develop practical processing and analysis systems.

- *intelligence systems* require that an analyst browse through billions of images looking for specific patterns only now recognizable to the human eye-brain system.

- *artificial Intelligence* applications will require new processors and data retrieval systems which will permit browsing through millions of documents or images to find relevant information stored in many computer systems and file systems. All of these systems will work across networks which permit them to be treated as one entity by the users.

Thus, several existing and emerging applications depend on network performance in the area of One Gigabit per second. What is required of a network to achieve the performance needed for the future generation of supercomputers and their applications?

Super Network Architectural Requirements

We have defined the need for a Super Network capable of carrying gigabits of data between many different high performance computers without error and for a diverse set of applications. What characteristics must a Super Network have? The following list characterizes the major components of the Super Network.

- The Data Links

    Communication links between the nodes of a Super Network must be able to carry data at rates greater than the desired throughput by some safe margin: at least 50%. Thus, for our Super Network, data links in the range of one to several **gigabits** per second are required.

- Connectivity to Hosts

    The host computers must be able to move data in and out of the computers at speeds approaching the network bandwidth. Therefore, for our Super Network, speeds of the host channel or external busses for the highest performance computers must approach one gigabit per second. The network should also be able to connect hosts from heterogeneous vendors to permit the addition of future processors. Channel and bus connections which are standards would improve the ability to connect with as many computers as possible.

- Network Processors

    The nodes of a network must be able to accept data from hosts and perform network processing at rates commensurate with the data links. Network processors in the Super Network must be able to process multiple thousands of packets per second (at least 25,000 - 50,000 per second) to ensure adequate throughput for both small and large files. Specialized network processors with custom hardware and software specifically designed for the network task may be required to achieve effective data rates approaching those of the data links.

- Protocols

  Protocols in a network ensure reliable data exchange between hosts and permit sharing of network resources. A Super Network must use an efficient protocol with little overhead in relationship to the amount of data normally being transmitted. This requires large packets of data to be exchanged to reduce overhead. Protocols recognized as Standard would be helpful to permit interoperability among heterogeneous vendors and to permit re-utilization of software. Standards today include protocols developed by the US Dept of Defense (TCP/IP) and the International Standards Organization (ISO TP4).

- Topology

  The method that we use to connect the hosts together must allow for maximum utilization of the network for certain applications. The topology must recognize that some applications will require dedicated bandwidth for periods of time. The topology should minimize collisions or contention for the network bandwidth to improve effective performance. Further, no restrictions should exist on the number of hosts that can be connected.

- Practical Considerations

  - Compatibility with existing network applications to minimize software efforts;

  - Support of a wide variety of operating systems;

  - Cost effectiveness;

  - Security capabilities to prevent unauthorized disclosure of data;

  - Long distance links to cover entire campus or corporate environments which may cover many miles.

  - Management facilities to allow reconfiguration when required, location of problems, tuning of network parameters to match usage environment, improve performance, and many other aspects of day to day maintenance of a large network environment;

  - Dynamic assignment of bandwidth to satisfy special uses of the network in the supercomputer environment.

What are the lessons that we can learn from the networking architectures and experiences of the past? A quick look at the following can help highlight new directions that might be explored for the Super Network.

Most high speed network architectures of today have been organized as either bus or ring topologies. Bus architectures generally connect all hosts on a single coaxial cable that is used in an arbitration mode. Collisions are usually detected and recovered from, or the cable is shared on a time demand basis. Hosts are connected to the bus or ring data links via network adapters or nodes which provide delivery of network packets between nodes with error detection and/or correction. Most high speed network products today provide what is known as the Physical Level and Data Link Level (Level 1 and Level 2 of the ISO model) service. The ability to move data reliability between adjacent nodes is provided in the network adapters. Higher level protocols resident in the host computer provide end to end reliable data transport known as the Transport Level. Applications which provide other services, such as File Transfer and Virtual Terminal Services, also reside in the host. Data is segmented into packets by the Transport software in the host to permit sharing of the network bandwidth by many processes and to reduce the amount of data to be resent when errors occur. Most of the protocols are designed to work over any quality data link with average efficiency.

For our Super Network, we must reduce the overhead inherent in today's networks to make the most out of the media bandwidth provided. Bottlenecks which cause such overhead should be explored further.

• Transport Protocol Processing

Transport level protocols take the responsibility for end to end reliable data transfer. An application calls a transport service interface routine to deliver it's entire buffer to a remote computer. This buffer may contain a graphics image, several megabytes of disk resident data, or field parameters for an entire iteration of a simulation ready for processing by a remote process. The Transport protocols interface with Network protocols and Data Link protocols responsible for routing the data in packet form to each network node over hardware dependent data links.

Host-resident Transport level protocols have several problems which affect the network performance:

   • the host processor must be shared with other real-time activities and user processes and thus compete with the network processing;

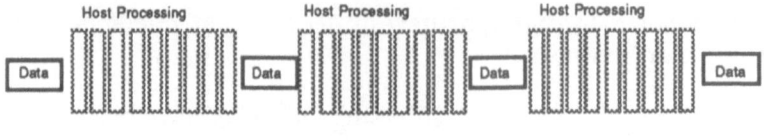

FIGURE 6.

- host computer time is consumed by networking mechanics. It must be spent processing headers/trailers for packetization, connection dialog, checksum computation, data movement, error recovery, and other network housekeeping chores;

- data is packetized in packets to permit shared use of networking facilities: each packet is normally an I/O operation for the host, thus creating multiple interrupts to the I/O system and perhaps the CPU for each buffer of user data;

- the computer architecture is not generally optimized for network processing.

These factors contribute to inefficient use of the network and host computer time. Inter-packet processing time is long relative to data transfer time. A several million bit per second network may rarely achieve more than 10% of its capacity (network bandwidth) from these host protocols and more typically in a loaded environment only achieve 1-3% of the capacity.

Many things can be done to alleviate this in the host, including increasing the packet sizes, dedicating host processors, improving the I/O path, and other techniques. However, another alternative is to place the major portion of the network processing in a separate processor designed and dedicated to network processing. In this alternative, the entire user's buffer is sent as a single I/O operation to the network processor. Host computer time is reduced, interrupts are dramatically minimized since packets are no longer configured in the host, and the separate processor can use specialized hardware to improve the network processing. Therefore, the profile in the host now looks more like the following figure:

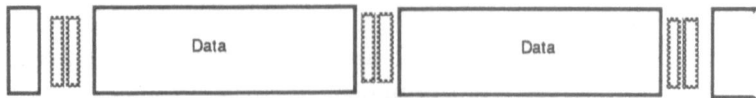

FIGURE 7.

Now, the network performance is more dependent upon factors than can be enhanced by unique hardware and software independent of the host architecture or performance.

Error strategies in a Super Network must not cause performance delays. Most network protocols have assumed that data links were generally unreliable. Fiber links and their components today have characteristics that provide error rates as low as 10^{-12} per second. This does not exempt us from otherwise providing error correction, but it does affect the strategy that can be used to improve performance. In the worst case, synchronous error strategies cause a full handshake between each host pair after each network packet. The Super Network should be able to support asynchronous data acknowledgement.

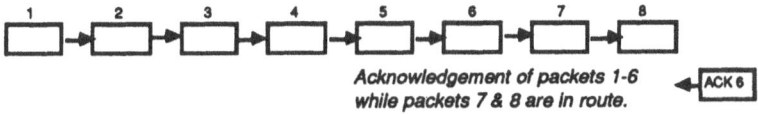

Acknowledgement of packets 1-6
while packets 7 & 8 are in route.

FIGURE 8.

Many packets should be sent before an acknowledgement is required. If data is missed, the data must be retransmitted from the point that it was last received properly. So we have traded brute performance for a longer error recovery period. This is possible with the lower error rates available on the data links today.

Frequently, high priority applications will require dedicated bandwidth from the Super Network. For instance, special equipment connected to the host may require dumping several megabytes of data in real-time. Making movies in real-time require that up to several seconds of dedicated network bandwidth be available to allow synchronized access to the graphics device. It is important that the Super Network provide such a dynamic method to permit dedicated usage without totally shutting down the network for extended periods or requiring any operator intervention.

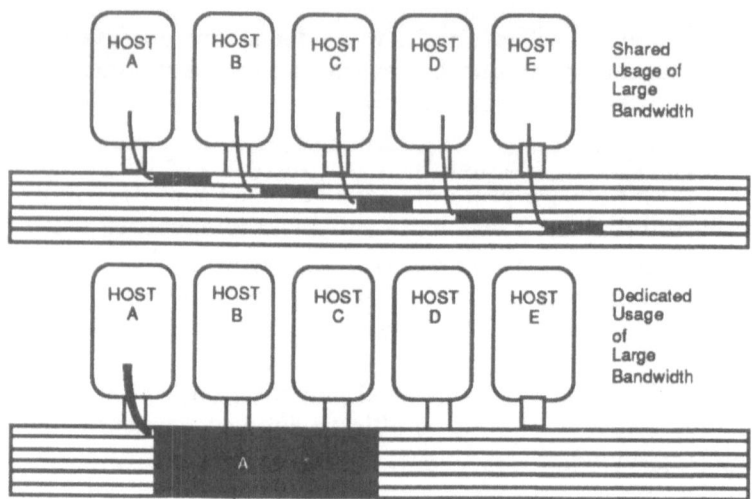

FIGURE 9.

The Super Network must also be able to support a large variety of processors. Frequently, a supercomputer is located at the top of a hierarchy of processors interconnected by many networks. The Super Network should provide gateways to many networks for connectivity, but at the same time should provide direct connection to as many processors as possible. For instance, although workstations might normally be connected via Ethernet or FDDI based networks due to cost considerations, it may be important for several applications running on the workstations to obtain as high a bandwidth as possible between the workstation and the supercomputer. For this reason, our Super Network should not be limited to just Supercomputers. Workstation connectivity is very important for the supercomputer user.

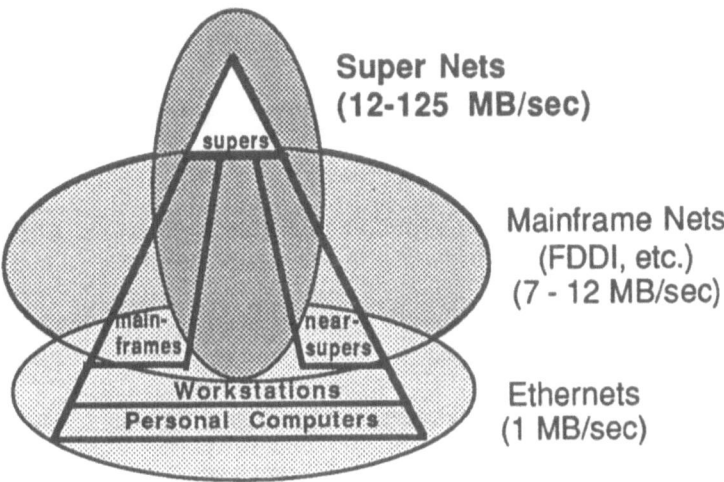

FIGURE 10. Hierarchy of computing.

• Topology

In a high speed network, where gigabit speeds are important, the network must provide the capability for achieving a large portion of the network bandwidth for a single application when required. Most networks today are a flat structure sharing a common bus which all applications compete for bandwidth equally.

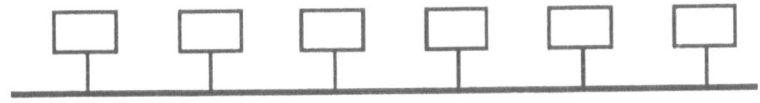

FIGURE 11.

Virtual terminal traffic, with byte-sized packets and very bursty traffic profiles compete with large file transfers consisting of several megabytes in size. In fact, most architectures require that all hosts compete in a contention based system for a common link resource: the cable bus. Ideally, the topology for the Super Network permits several independent non-contending busses, thus grouping local traffic together, minimizing contention across the entire network.

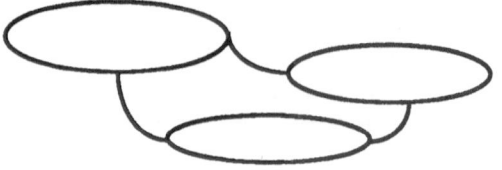

FIGURE 13.

Localizing high bandwidth traffic into non-contending groups of hosts can increase the bandwidth of the entire network.

In summary, the Super Network must address the following aspects related to network processing in a supercomputer environment:

- Specialized Network Processors for Transport Level Interface

- Support of gigabit/sec data links and host connections

- Low error rate data links

- Topology which permits segregation of localized traffic

- Asynchronous error strategies for better protocol performance

- High performance hardware

- Strategies for dynamic allocation of bandwidth to specific hosts or of applications

- Support a large number of diverse processors

Future Prospects for Super Networks

Today, several networks support the scientific and engineering computing community. The networks range from 1 to 100 million bits per second in their total bandwidth and from .2 to 8 million bits per second in effective throughput. None of these is adequate for the requirements stated above. The fact that so many computer vendors now offer computers for the science and engineering marketplace which produce results in the several million to billion operations per second range tell us that the need that we have established for a Super Network is real today. In the next five years, it will be overwhelming.

Several standards organizations have taken note of these requirements, but none have completed standards for such high speed interfaces. The ANSI X3T9.4 committee has developed the FDDI (Fiber Distributed Data Interface) fiber optic ring network capable of data link performance up to 100 million bits per second. Many vendors will be offering FDDI interfaces within the next year

or two.   However, the FDDI is only a data link definition, and
without strategies as outlined above for network processing,   the
effective data rates will remain in the 5 - 15 million bits per second
range.

The ANSI X3T9.3 committee has developed a draft standard for a
High Speed Channel interface to computers that support either 100 or
200 Million Bytes per second between two hosts.   This interface
standard is promising to provide new incentives for network and
computer vendors to develop the Super Networks described above.

UltraNet™ - One Super Network

At Ultra Network Technologies, we have developed a network
product which we think meets most of the needs of the Super Network
described above.

UltraNet® communicates between hosts at rates up to ONE
GIGABIT PER SECOND. The Ultra Network consists of:

• Host Adapters which connect to a variety of hosts and provide a
Transport Level Service to the computer host, thus permitting large
data blocks to be output directly from the user's buffer to the network
processor.   A combination of special hardware and software in the
Adapter permit a very high EFFECTIVE rate of transfer between
hosts on the Ultra Network. The Adapter actually utilizes standard
protocols (ISO TP4) for end-to-end host data transfers. Network
bandwidth can be dynamically assigned to hosts for dedicated
applications.

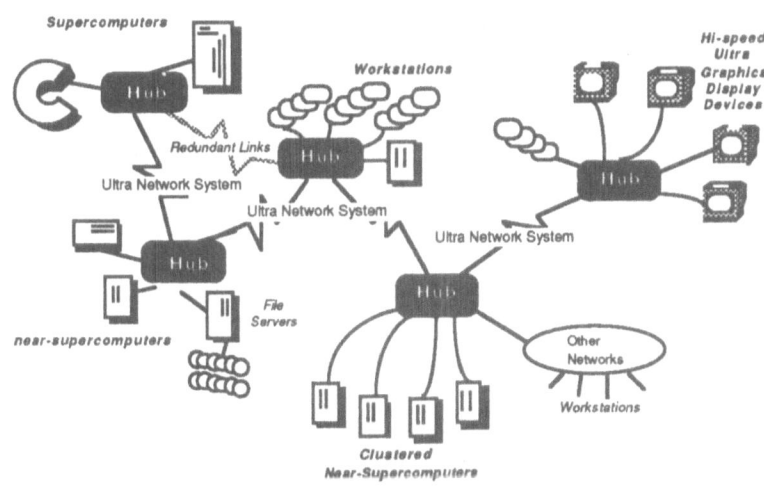

FIGURE 12.

• UltraBus™ data link is a very fast data bus backplane that supports several Ultra Network Adapters at over a GIGABIT/SEC in a single Ultra Network Hub.

• UltraNet 1000 Hub permits several network adapters to be grouped for low contention/high performance data transfer for the highest performance hosts (up to 800 Mbit/sec). This Hub can connect several lower performance hubs or additional UltraNet 1000 hubs to produce a large UltraNet environment. A separate Network Manager processor is provided which permits control of network resources by the site network administrator.

• UltraNet 250 Hub supports multiple lower performance processors at high network speeds and permit cluster's of these connected to an UltraNet 1000 or in a stand alone environment at speeds up to 250 M bits/second.

• Hub to Hub Link Adapters permits data transfer between Hub's at speeds up to a GIGABIT/SEC. The Link Adapters are invisible to users in a large network of interlinked Hubs. Initially, Hubs can be located up to 4 kilometers apart over multiple fiber optic pairs depending upon transfer rate.

• Hub to Host Link Adapters permit mutliple individual hosts (or workstations) to be directly connected to the UltraNet each sharing up to 250 million bits per second. Each host can be up to 4 kilometers from a Hub.

• Ultra Host Software provides familiar interfaces to the users, including a Sockets Interface to the UltraNet Transport Service; file transfer applications including FTP, rcp and FTAM; and a Sockets Network Driver for internetting between other networks and UltraNet using TCP/IP or other protocols. Operating Systems supported include UNIX™, UNICOS™, Concentrix™ and CTSS. Support for IBM's MVS and VM is also available.

• Ultra Graphics Display Device is a networked high performance, high resolution framebuffer. Users on any host in the network can display images up to 1280 x 1024 x 24 bits in color at speeds up to 24 frames per second animation. Higher animation rates are possible with lower resolution.

Future products under development include:

• Gateway and Router Adapters will permit interfaces to other networks for "backbone" applications for FDDI, and Ethernet networks.

• <u>Wide Area Link Adapters</u> will permit interfaces to multiple T1 (1.3 million bits/sec) and T3 (44 million bits/sec) data communications services for linking remote UltraNet systems.

The current list of host interfaces includes a sample of those systems used today in the science and engineering world. They include:

- Cray XMP & YMP (HSX Channel at 100 MBytes/sec)

- Cray 2 (HSX Channel at 100 MBytes/sec)

- Alliant FX-8/80 Interface

- Convex C-1& C2 Interface

- Sun Microsystems 3 and 4 Interface

- Silicon Graphics 4D & Power Series Interface

- IBM 3090 (HSC Interface at 100 MBytes/sec)

- IBM 3090, 308* and 43** (BMC Interface in Late 89)

Several other interfaces are in development. Emerging standards, such as the VMEbus, ANSI X3T9.3 High Speed Channel, and FDDI will permit the Ultra Network to interface to a variety of hosts as vendor's develop these standard interfaces.

Supercomputers, Super Workstations and Super Networks will offer scientists powerful tools for discovery through interactivity and visualization of results as they have never known before. Tools which provide order-of -magnitude differences in capabilities over past tools are sure to offer new insights into scientific discoveries.

UNIX™ is a trademark of AT&T, UltraBus™ and UltraNet® are trademarks of Ultra Network Technologies; UNICOS, HSX, Cray XMP,YMP and Cray 2 are trademarks of Cray Research, Inc. Concentrix™ is a trademark of Alliant Computer Systems.

Cellular Array Processor CAP and Visualization

Hiroyuki Sato, Mitsuo Ishii, Morio Ikesaka,
Kouichi Murakami and Hiroaki Ishihata

Fujitsu Laboratories, Ltd., Kawasaki
1015 Kamikodanaka, Nakahara-ku, Kawasaki 211, Japan

ABSTRACT

The general-purpose cellular array processor (CAP) we developed features multiple-instruction stream multiple-data stream (MIMD) processing and image display. Several hundreds of processor elements can be connected together. The present system uses 256 processors. Each processor element consists of a general-purpose microprocessor, memory, and a special VLSI chip that performs parallel-processing-specific functions such as processor communication and synchronization. The VLSI has two 2M byte/s independent common interfaces for data broadcasting and six 15M bit/s serial communication ports for local data communication. The chip can also process image data in real time for multiple processors.

CAP has been successfully applied to visualization applications such as ray tracing, three dimensional reconstruction of CT slice images, and realtime visualization of numerical simulation.

1. INTRODUCTION

Parallel processing is a key to speeding up problem solving. Advances in VLSI technology have made available powerful microprocessors and large-capacity RAM chips, enabling us to construct parallel processors consisting of several hundreds of processor elements. Examples include the iPSC[1], CM-1[2] and N-CUBE[3]; some of these are now being marketed. Very few cost-effective uses have been found for such systems, however. Much more research is needed on applications for parallel computers.

The cellular array processor CAP is a parallel computer we developed to study parallel processing, from hardware architecture to application software. CAP is based on the following design principles:

[1] MIMD Processing Using Highly Intelligent Processor Elements

For general-purpose use, each cell must provide a multitasking environment to enable the system to be applied easily to the wide range of problems that can be solved using parallel processing. It should also be capable of being used as a tool for developing and evaluating parallel algorithms.

[2] Intercell Communication

The architecture must allow a system of several hundreds of cells to be constructed. Using shared memory to construct such a system is needlessly complicated, so we used communication ports to communicate between processors. Hypercube connections are effective for random communication between processors, but do not work well for global communication, such as data broadcasting from one processor to all other processors, or for collecting data from all other processors. Hardware support for these types of communication is needed to reduce communication overhead.

No single network topology suits all applications, which means that networks of different topologies should be used depending on the application.

[3]   Realtime Image Display

Graphics is essential to a good user interface. We incorporated a realtime image display that makes CAP well suited to applications involving image data, such as computer graphics and image processing. One of CAP's best applications has been in image generation[4][5].

Graphics is also useful when debugging software. The behavior of multiple processors during execution can be much more easily understood with a realtime display of each processor's status.

2. CAP ARCHITECTURE

2.1. Overview

Figure 1 gives the standard CAP-C5[9] hardware configuration of 256 processor cells in a two-dimensional array. Figure 2 shows the CAP-C5 hardware. Each cell is connected to four adjacent cells. Cells at the boundaries are connected to cells at the opposite sides. The network topology resembles the surface of a torus. In addition to intercell connections, the common command bus links all cells to the host computer, and the common video bus links all cells to the video interface.

2.2. VLSI Architecture for Parallel Processing

The CAP-VLSI chip[6] is a key processor component. It operates together with the microprocessor and memory elements. The chip has the following functions:

[1] Window controller for fast image-data transfer

[2] Two common bus interfaces for global communication

[3] Hardware synchronizer

[4] Six intercell communication ports configuring an intercell communication network

Figure 3 shows the cell hardware configuration. Table 1 lists cell specifications. The CAP-VLSI chip (Figure 4) is fabricated using a channelless CMOS gate array chip[7]. Table 2 lists chip specifications.

2.3. Window Controller

Image generation is speeded up by partitioning an image area into noncontinuous subareas and using a large number of cells to generate subimages in parallel. Each cell's window controller maps its subimage data onto a part of the screen. A frame of image data, partitioned and distributed to many cells, is reconstructed as one complete image by cell window controllers. The mapping pattern can be changed as needed. Each cell maps its image data onto dispersed block areas (Figure 5).

Changing the size and intervals of blocks produces a variety of patterns (Figure 6), and makes it easy to implement different image-generation algorithms. It also helps even out load distributions, as explained later.

Each window controller performs such tasks as address generation and arbitration of image memory access from the video bus and MPU. Image data is read via the video bus every refresh cycle (33 ms) for refresh display. Image memory can be used both for full color (8 bits each for red, green, and blue) and monochrome (8 bits per pixel). Images can be input and output simultaneously in real-time by using separate video buses for input and for output. Image data is easily input from a TV camera or other device.

2.4. Common Bus Interface

The host computer broadcasts to all cells via the common command bus. The CAP-VLSI chip has two command bus interfaces, one for the host and the other for cells, whose use enables a variety of hierarchies to be configured (Figures 7 (a) and (b)).

The command bus interface has 16-bit 8-word first in, first out (FIFO) memory. This helps to reduce differences in cell-processing time. A 3-wire handshake protocol is used in 1 to N communication. Any cell can broadcast data to other cells and to the host computer, although arbitration is needed when more than one cell requests to broadcast at the same time.

The command bus interface also controls resets, interrupts and bus arbitration. The host computer or host cell can reset or interrupt all slave cells or any one cell. The CAP-VLSI chip uses hardware polling to arbitrate bus requests and to specify individual cells. Each cell has a two dimensional address and polling is executed in 2 phases. The first polling determines the column position of the requesting cell, then the second polling determines the row position of the requesting cell. To poll cells sequentially, the time increases in proportion to the number of cells. Our polling technique reduced the polling time from $O(N)$ to $O(\sqrt{N})$, where N is the number of cells.

2.5. Hardware Synchronization

Global synchronization or synchronization among some cells is needed to execute most of the parallel algorithms. Each cell has status registers for synchronization. Outputs of the status register of each cell are wire-ORed, and the line status can be read by the host or by any cell.

Cells can set status to indicate completion of a process or other states. The host or any cell can read the logical OR (or AND) statuses of all cells to detect the completion of processes in a number of cells.

Each status line can handle a single independent synchronization request. Conventionally, the more the synchronization requests, the more the status lines. Multiplexed synchronization control (MSC) avoids this situation without increasing the number of status lines. MSC uses two control lines to automatically synchronize 16 synchronization requests.

2.6. Intercell Communication Ports

Each cell communicates with adjacent cells via six full-duplex serial communication ports. Networks of different topologies, such as two-dimensional cell arrays, three-dimensional cell arrays, and hypercubes of up to six dimensions, can be configured (Figures 7 (c) and (d)).

The CAP-VLSI chip uses a bypass for intercell communication. Any two of six ports can be connected directly by commands from the MPU, enabling fast communication between distant cells by cell input and output port connection on the route. Paths are set and released dynamically.

Data is transferred in 19-bit packets -- 16 data bits and 3 header bits to identify the packet type (data, interrupt, nonmaskable interrupt, or acknowledge). Connected cells communicate by a handshake protocol using data and acknowledge packets. Using interrupt packets, a cell can interrupt another cell's MPU whenever the cell is not bypassed. Nonmaskable interrupt packets are used to interrupt bypassed cells.

3. OPERATING SYSTEM

Problems executed by CAP are divided into parts that can be processed in parallel, then mapped to cells, which communicate by exchanging messages.

Figure 8 shows software configuration.

3.1. Cell OS

We developed a cell OS that controls task execution in each cell and supports intertask communication. The cell OS has several advantages. First, having programming tasks as elements to be executed in parallel by many processors or concurrently in one processor makes it easier to extract parallelism in problems, making programs more efficient. Second, application programmers are freed from designing complicated procedure scheduling and can develop programs more easily.

Messages are usually sent in packets. Messages from other tasks are queued and read in the sequence they arrive. Message destinations are specified by cell and task numbers. Messages sent with only a task number are broadcast to destination tasks in all cells. Communication tasks pass messages between tasks.

The cell OS supports three-dimensional mesh, hypercube, and hierarchical connections. VLSI bypassing provides application programmers with two types of bypassed communication. One is static, in which the host broadcasts bypass information to cells to establish fixed connections between specified cell pairs. The other is dynamic, in which a cell sends a message directly to the destination cell, dynamically bypassing cells on the route.

3.2. Cell Driver

The cell driver, which resides in the host, dynamically allocates tasks to cells, and broadcasts data from an application task in the host to cells and collects data from cells.

3.3. Display Manager

The display manager, a basic software package for image generation and display, provides standard patterns for mapping cell subimages onto the screen (Figure 6).

Two-dimensional primitives can also drawn, and animated display of multiple image frames is very useful in certain applications.

4. APPLICATION TO VISUALIZATION

4.1. Parallel processing in image generation

Image generation by CAP is performed in two stages (Figure 9). In the first stage, object or three dimensional space is divided and distributed to many cells and is processed in parallel. In the second stage, the image space is divided into subareas and is processed in parallel.

Both object and image are divided for parallel processing (Figure 10). In ray tracing, however, the object space is not divided. Each˜cell has the same copy of the entire model data. Image space is divided in the dot mode. In scanline or z-buffer algorithms, the object is divided to subsets of primitives such as polygons and distributed to cells. After viewing transformation and polygon rendering are performed in parallel, generated subimages with z values are redistributed to cells in which image space is divided in the line mode. Hidden surface removal by a z-buffer algorithm is then performed. In volume rendering, three dimensional cubic space is divided and is processed in parallel.

4.2. Ray tracing

The ray tracing algorithm[8] simulates optical phenomena such as reflection, shadows, translucency, and refraction. It generates quality images, although this requires a large amount of calculation.

4.2.1. Parallel Ray Tracing

Rays can be traced pixel by pixel. Ray tracing can be speeded up by dividing the screen area and having multiple processors process each small area in parallel (Figure 11). We implemented a ray tracing program into CAP systems[5].

4.2.2. Static Load Distribution

To enable performance to improve with the number of processors, it is very important to distribute the calculation load evenly. Because rays can be processed independently, load is evenly distributed using the dot the mode (Figure 6), in which each cell takes charge of a similar proportion of heavy-load and light-load pixels.

Each cell has a copy of all model data, together with camera and lighting information.

4.2.3. Antialiasing Process Using Inter-cell Communication

Because of ray tracing's point-sampling nature, undesirable effects called aliases are generated in synthesized images, examples are the staircase pattern along straight edges, the moire patterns in finely textured areas. To reduce these aliasing effects, antialiasing is performed as needed. To calculate a pixel's antialiased intensity, intensities of the four adjacent pixels must be known.

Rays are traced in dot mode, in which adjacent cells process adjacent pixels on the screen, e.g.) adjacent pixels to the right neighbor are in the right-hand cell, and those to the left are in the left-hand cell. Thus, each cell needs only to communicate with the neighboring cells to collect other pixels' intensities. The communication time for local data transfer is minimal.

4.2.4. Experimental results of ray tracing

Experiments were performed using the 64-processor CAP-C3 and 256-processor CAP-C5. CAP-C3 processors (cells) are similar to those of CAP-C5, except that VLSI chip functions are implemented by discrete hardware. Figures 12 and 13 are examples of ray-traced images. The results of experiments showed that cell processing times differed a maximum of 10 percent when 256 cells were used. This evenness of load distribution results in a linear performance improvement that increases with the number of cells (Figure 14). Ray tracing with 256 cells is over three times faster than Fujitsu's M-380 mainframe (Table 3).

Image quality is improved by antialiasing, with additional calculation time for adaptive oversampling. During oversampling, multiple rays are traced on a pixel region, and subpixel intensities are averaged (Table 4). Communication overhead for antialiasing is small compared to the total processing time.

4.3. Three dimensional reconstruction of CT slice data

CT (Computer Tomography) scanners are used to produce slice images of objects. Three dimensional reconstruction of the original object from sliced images is useful for various areas.

4.3.1. Slice image buffer

Slice images translated to the eye coordinate system are stored in cells. Each slice is divided in the line mode. This data structure is called a slice image buffer (Figure 15). First, original slice images are broadcast from the host. Each cell receives slice images and translates them to the eye coordinate system. Each cell stores only data for assigned scanlines to the slice image buffer.

We implemented display function such as thresholding, hidden surface removal, pseudocoloring and cross sectioning. These functions use this slice image buffer.

The basic display algorithm is shown below in pseudo C code.

```
clear video memory
for (scanlines assigned to this cell){
  for (pixels on the scanline){
    for (slices from front to back){
      if (the pixel is not in the cut volume and the CT value is in the range specified){
        write lookup table color value corresponding to the CT value
        break
      }
    }
  }
}
```

Figure 16 shows a display example of the upper part of a human brain. Twelve slices are used. The resolution of the original slice is 512 by 512 pixels. Each pixel consists of 8 bits. The screen resolution is 512 by 384 pixels. The time required to generate a frame from the slice image buffer is about three seconds when 64 cells are used.

Rotation in an arbitrary direction requires remaking of the slice image buffer. However, rotation in the horizontal direction (around the vertical axis) can be done quickly without remaking the slice image buffer. Horizontal rotation is used for animation display. Multiple frames are made and stored in the video memory and then displayed repeatedly in an animated sequence. Animation display is effective in understanding the three dimensional structure of the object.

Color Insert

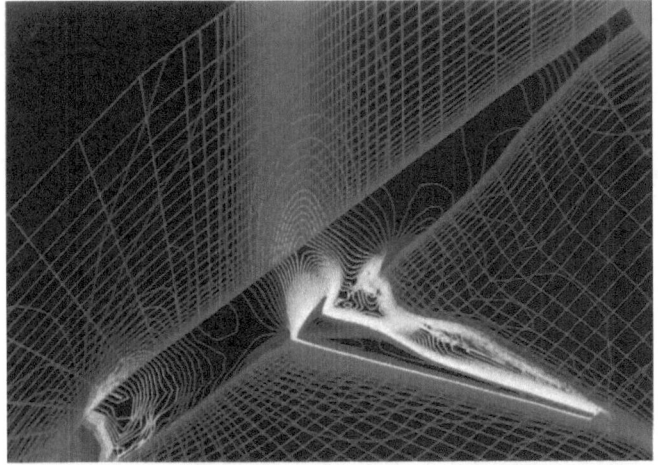

FIGURE 1

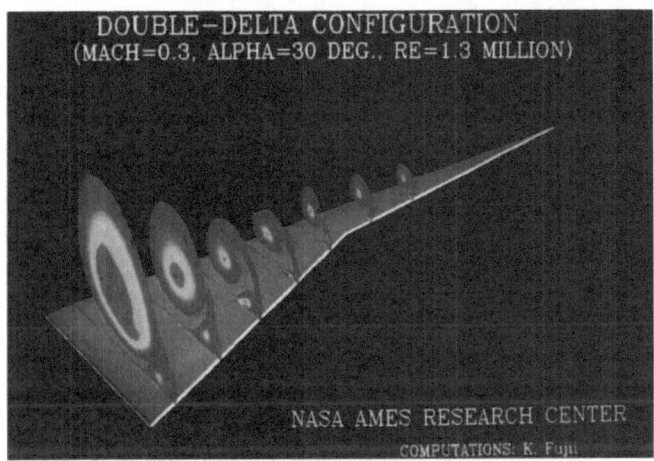

FIGURE 3

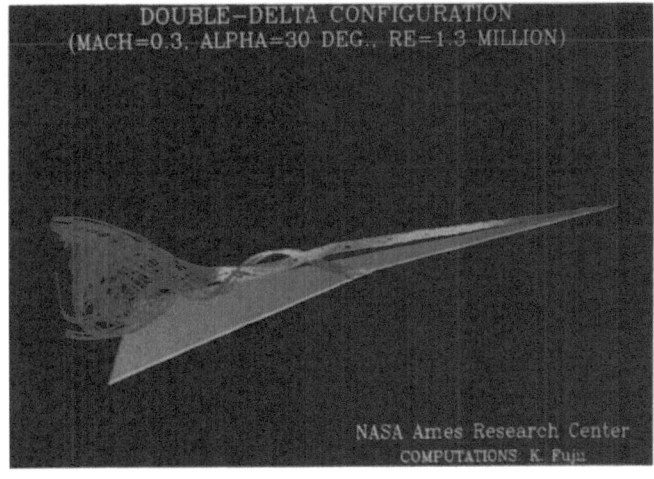

FIGURE 4

Figures 1, 3, and 4 from "Supercomputers and Workstations in Fluid Dynamics Research," by Kozo Fujii, pp. 11 and 12.

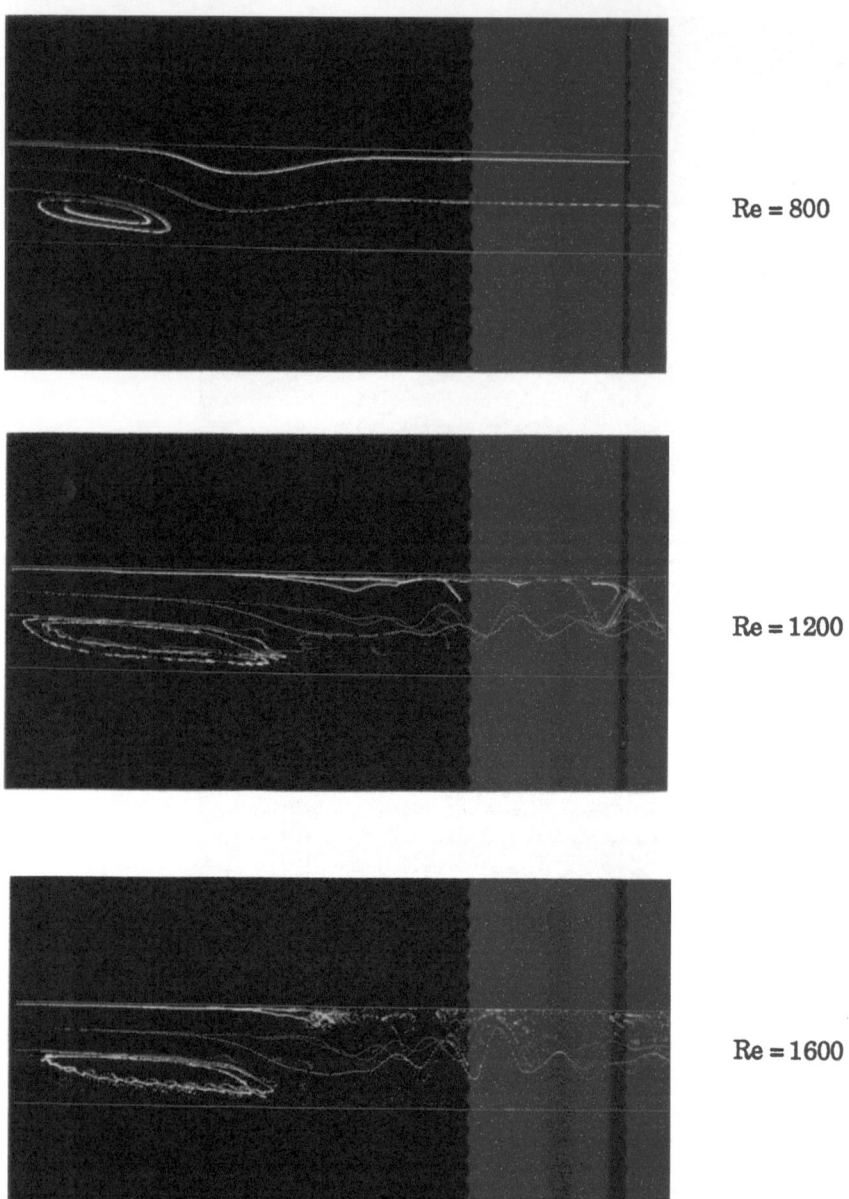

Re = 800

Re = 1200

Re = 1600

Figure 7a from "Numerical Simulation of a 3-D Backward-Facing Step Flow," by Hiroshi Takeda and Erika Misaki, p. 23.

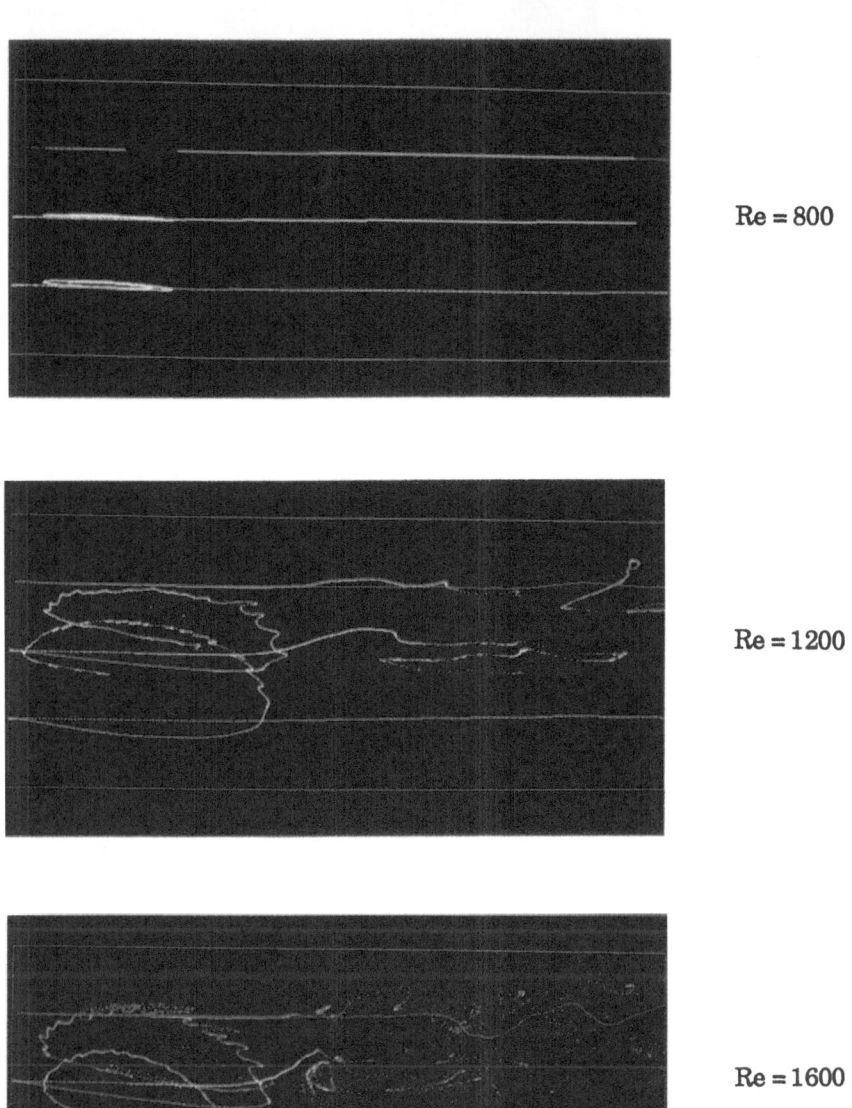

Re = 800

Re = 1200

Re = 1600

Figure 7b from "Numerical Simulation of a 3-D Backward-Facing Step Flow," by Hiroshi Takeda and Erika Misaki, p. 24.

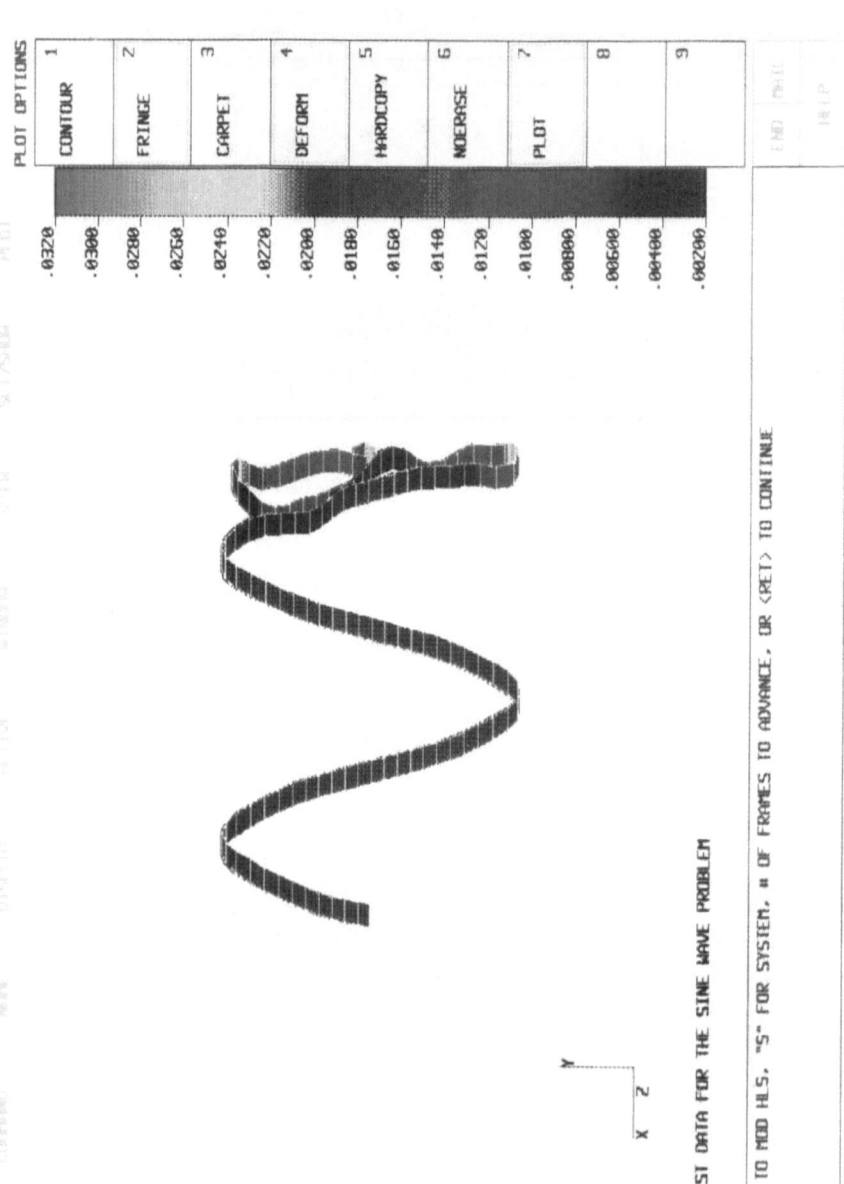

Figure 1 from "System Solutions for Visualization: A Case Study," by Kohei Kumazawa and Christopher Eoyang, p. 29.

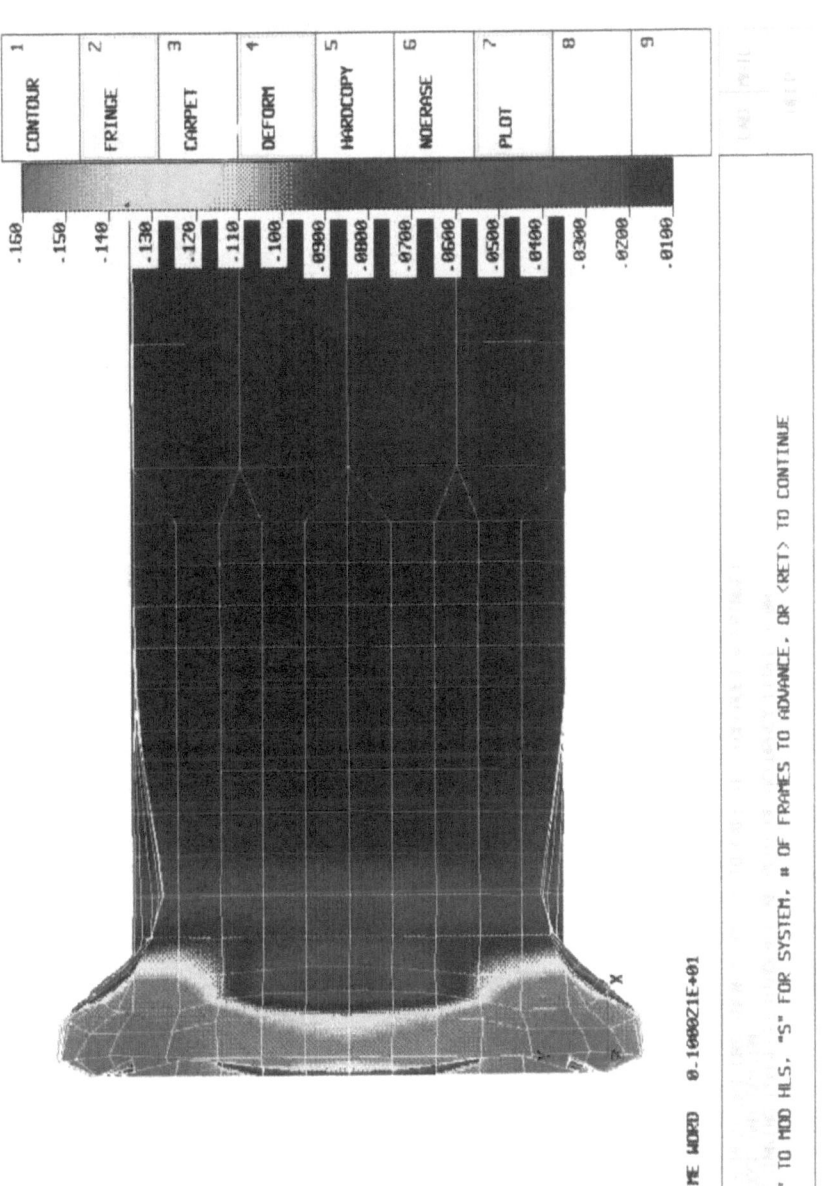

Figure 2 from "System Solutions for Visualization: A Case Study," by Kohei Kumazawa and Christopher Eoyang, p. 30.

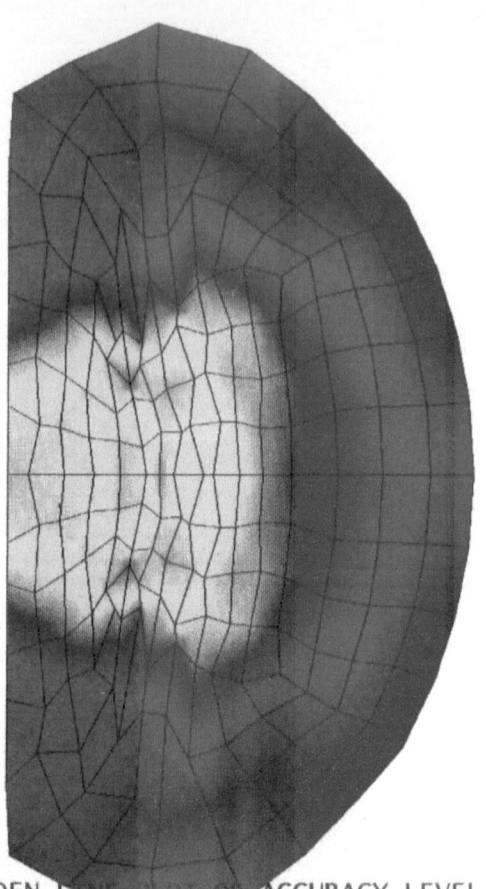

BEGINNING PHASE-II HIDDEN LINE PLOT OR ACCURACY LEVEL 1.00
LOCAL SHADING: D=DITHER, L=LABELS, O=OUTLINING, E=END
>

Figure 3 from "System Solutions for Visualization: A Case Study," by Kohei
Kumazawa and Christopher Eoyang, p. 31.

Figure 12 from "Cellular Array Processor CAP and Visualization," by Hiroyuki Sato, Mitsuo Ishii, Morio Ikesaka, Kouichi Murakami, and Hiroaki Ishihata, p. 112.

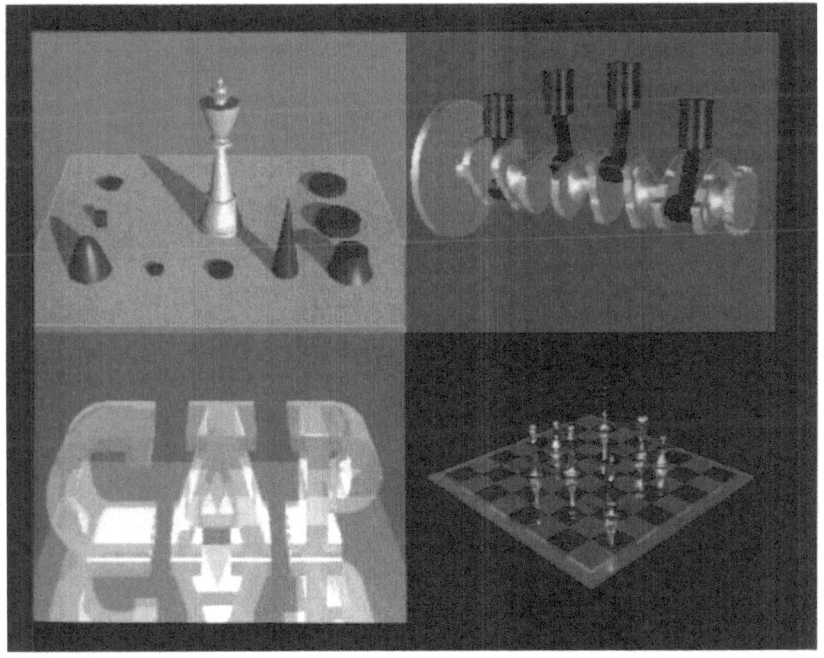

Figure 13 from "Cellular Array Processor CAP and Visualization," by Hiroyuki Sato, Mitsuo Ishii, Morio Ikesaka, Kouichi Murakami, and Hiroaki Ishihata, p. 113.

Figure 16 from "Cellular Array Processor CAP and Visualization," by Hiroyuki Sato, Mitsuo Ishii, Morio Ikesaka, Kouichi Murakami, and Hiroaki Ishihata, p. 114.

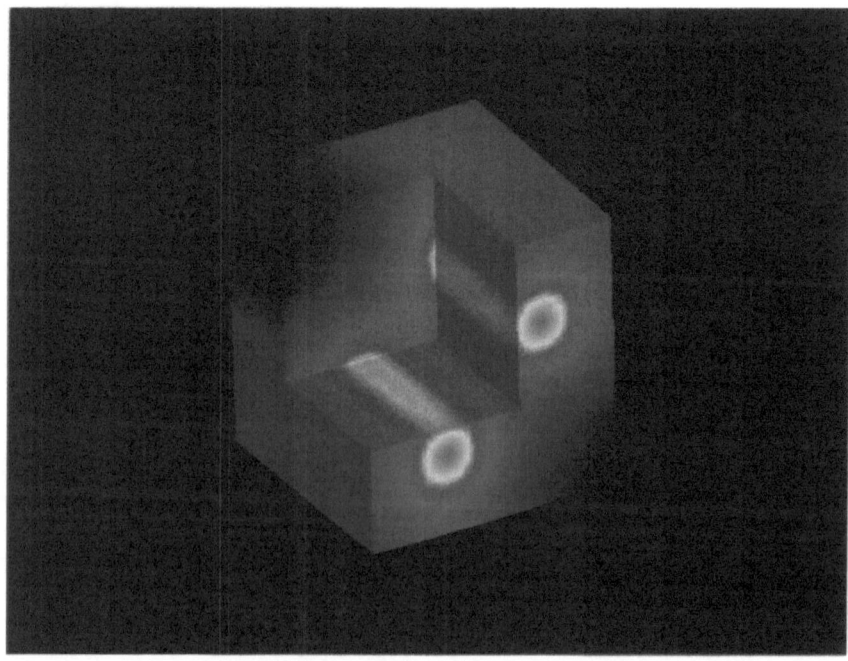

Figure 19 from "Cellular Array Processor CAP and Visualization," by Hiroyuki Sato, Mitsuo Ishii, Morio Ikesaka, Kouichi Murakami, and Hiroaki Ishihata, p. 116.

(a)

(b)

Figures 4a and b from "Lighting Simulation," by Eihachiro Nakamae, p. 160.

(a)

(b)

Figures 6a and b from "Lighting Simulation," by Eihachiro Nakamae, p. 161.

(a)

Figures 8a and b from "Lighting Simulation," by Eihachiro Nakamae, p. 163.

(a)

(b)

Figures 10a and b from "Lighting Simulation," by Eihachiro Nakamae, p. 164.

(c)

(d)

Figures 10c and d from "Lighting Simulation," by Eihachiro Nakamae, p. 164.

4.4. Realtime visualization of numerical simulation

As computer power increases, the visualization of computed results becomes more and more important. There are two problems in scientific visualization. One is the huge amount of the data to be handled. The best way to deal with it is realtime visualization of the simulation. No data need to be kept for post processes. The other is the communication bottleneck between computing and visualization systems. If computing and visualization are executed in the same system, the I/O bottleneck is easily eliminated.

4.4.1. Heat flow simulation using the finite difference method

We have developed an example program, which performs heat flow simulation in a solid material and visualizes the temperature distribution at the same time.

The basic equations are those for heat dissipation (Figure 17). These are solved by the finite difference method. In two-dimensional simulation, a five point stencil is used. The physical space is divided into rectangle areas and processed in parallel. In three-dimensional simulation, a seven point stencil is used. The physical space is divided into cubes and processed in parallel. We used an explicit calculation scheme called red and black SOR (successive over relaxation)[10]. To calculate next iteration step grid temperature values, only nearby gird temperature values are necessary. The temperature values of grids located at cell boundaries are exchanged through intercell communication lines.

4.4.2. Cross section display of volume data

In two dimensional simulation, it is easy to display the temperature distribution. In three dimensional simulation, volume rendering techniques are necessary. We implemented a cross section display function. Each cell handles a cubic volume. As mentioned earlier, two stage processing is performed to generate cross-section images of the volume. In the first stage, each cell renders cross section areas if any. The generated subimage data is then distributed to the cells. The image is divided in the line mode. In Figure 18, for example, the cell which handles the shaded cube renders these shaded polygons. Then, that cell sends these subimages to other cells line by line using intercell communication lines. The z value of each pixel is also sent for the hidden surface removal.

The cross section display of a temperature field (Figure 19) took about five seconds when 64 cells were used. It is possible to record the image on the video tape frame by frame to make an animation of the simulation.

Other volume rendering techniques such as isovalue surface display and projection are left for future study.

5. CONCLUSION

We have explained the architecture of the cellular array processor (CAP), which used specially designed CAP-VLSI chips to manage several hundreds of processor cells. CAP uses two independent common bus interfaces for data broadcasting and six serial communication ports for local data communication. The chip also has realtime image-data handling capabilities. Using CAP-VLSI chips and general-purpose microprocessors, we constructed a system with 256 processor elements.

CAP has been applied to visualization such as ray tracing, three dimensional reconstruction of CT slice images and visual simulation. In ray tracing, CAP generates high quality images efficiently by evenly distributing the calculation load using the dot mode. In three-dimensional reconstruction of the CT slice images, interactive handling and display is possible using the slice image buffer data structure in the line mode. We made a prototype program which performs realtime visualization of heat flow simulation. Combining the computing power of a massively parallel system and image generation functions will make it possible to construct an interactive visual simulation system.

6. REFERENCES

[1] Intel Corp., data sheets.
[2] W. Daniel Hillis, *The Connection Machine*, Cambridge, Mass., MIT Press, 1985.

[3] John D. Hayes, *et al.*, "Architecture of a Hypercube Supercomputer," *Proc. Int'l Conf. on Parallel Processing*, pp. 653-660., 1986.

[4] H. Sato, M. Ishii, *et al.*, "Fast Image Generation of Constructive Solid Geometry Using a Cellular Array Processor," *Computer Graphics*, 22(2), July 1985, pp. 95-102

[5] K. Murakami, H. Sato, *et al.*, "Ray Tracing Using Cellular Array Processor CAP"(in Japanese), *Information Processing Technical Report*, Vol. 86, No. 43, CAD-22-2, July 1986.

[6] H. Ishihata, M. Ishii, *et al.*, "VLSI for the Cellular Array Processor," *Proc. Int'l Conf. on Computer Design*, pp. 320-323., Oct. 1987

[7] H. Takahashi, *et al.*, "A 240K Transistor CMOS Array with Flexible Allocation of Memory and Channels," *IEEE Journal of Solid State Circuits*, Vol. SC-20, No. 5, pp. 1012-1017, Oct. 1985.

[8] T. Whitted, "An Improved Illumination Model for Shaded Display," *Comm. ACM* Vol. 23, No. 6, 1980, pp. 343-349.

[9] M. Ishii, H. Sato, *et al.*, "Cellular Array Processor CAP and Applications," *Proc. Int'l Conf. on Systolic Arrays*, pp. 535-544, 1988

[10] L. Adams and J. Ortega, "A Multi-Color SOR Method for Parallel Computation," *Proc. Int'l Conf. on Parallel Processing*, pp. 53-56, 1982

Table 1 Cell specifications

MPU	i80186 + i8087
RAM	2M bytes(with ECC)
ROM	64K bytes
Image memory	96K bytes
Video clock	15 MHz maximum
Common bus transfer rate	2M byte/s
Intercell transmission speed	15M bit/s

Table 2 Gate array specifications

Design rule	CMOS 1.8 μm
Chip size	13 mm square
Gate delay	1.5 ns/gate
Basic cell configuration	94 × 311
Package	256-pin PGA
I/O	220 lines, TTL-compatible

Table 3 Results of ray tracing experiments

Model (#Object)	Image generation times(s)		Performance rate
Conditions	CAP	FACOM M-380	CAP/M-380
BALL(125) Reflection	19	60	3.16
PISTON(179)	10	33	3.36
CHESS(144) Reflection Shadow	40	128	3.20

Display resolution: 512 x 384 pixels
Number of cells: 256 (CAP-C5)
(The FACOM M-380 has the equivalent processing power of the IBM 3081.)

Table 4 Timing results of anti-aliasing for the model PIANO

Test item	No anti-aliasing	Level 0 (Averaging four points)	Level 1 (Adaptive oversampling)
Communication time(s)	.0	2.0	2.0
Time of slowest cell (s)	56.6	68.1	121.1
Time of fastest cell (s)	55.7	67.3	104.8
Maximum deviation (%)	1.6	1.1	13.5

Maximum deviation = {(Time of slowest cell) - (Time of fastest cell)}
    /(Time of slowest cell) * 100
Number of cells: 64(CAP-C3)

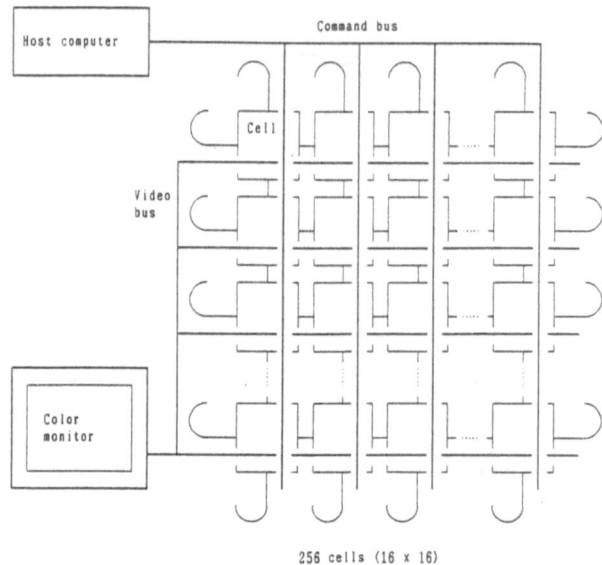

Figure 1 CAP hardware configuration

Figure 2 CAP-C5 hardware

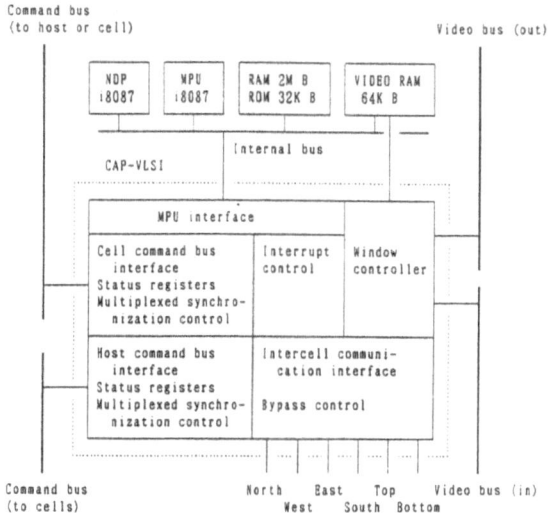

Figure 3 Cell hardware configuration

Figure 4 CAP-VLSI chip

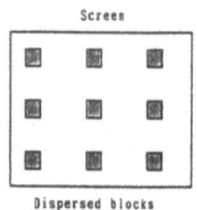

Figure 5 Subimage mapping

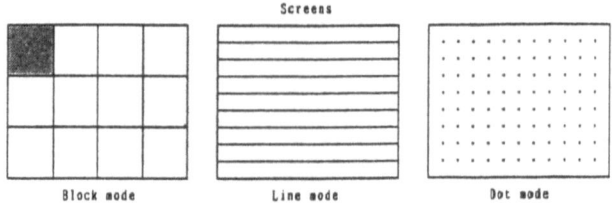

Figure 6 Mapping patterns

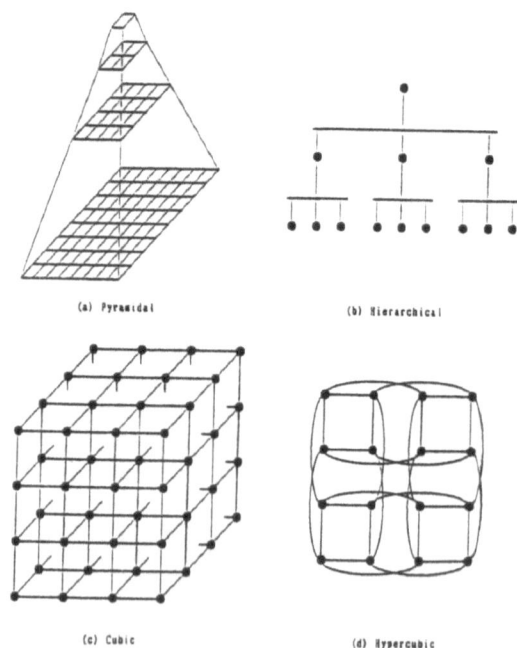

Figure 7 Network configurations

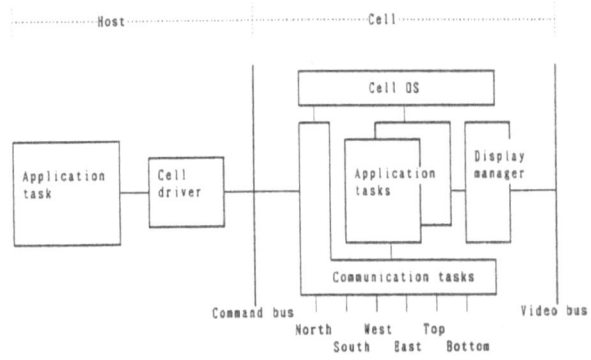

Figure 8 Software configuration

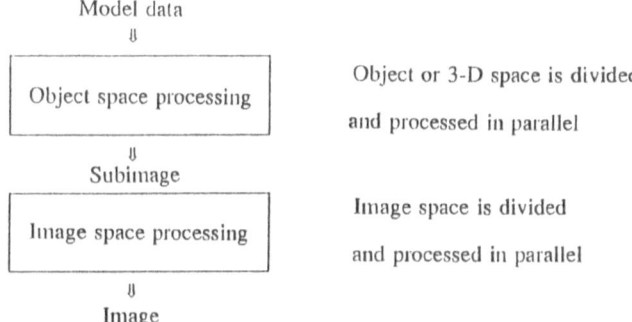

Figure 9 Image generation by CAP

	Algorithm		
Space	Ray tracing	Scanline/Z-buffer	Volume rendering
Object space	no division	Object division (polygons)	space division (Subvolume)
Image space	Division in dot mode	Division in line mode	Division in line mode

Figure 10 Space division for parallel processing

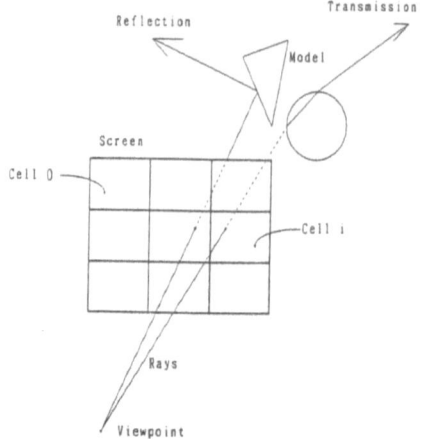

Figure 11 Parallel processing of ray tracing (block mode)

Figure 12 Ray-traced image of a chess board (color art for this figure may be seen in the color insert)

Figure 13 Ray-traced images of four models (color art for this figure may be seen in the color insert)

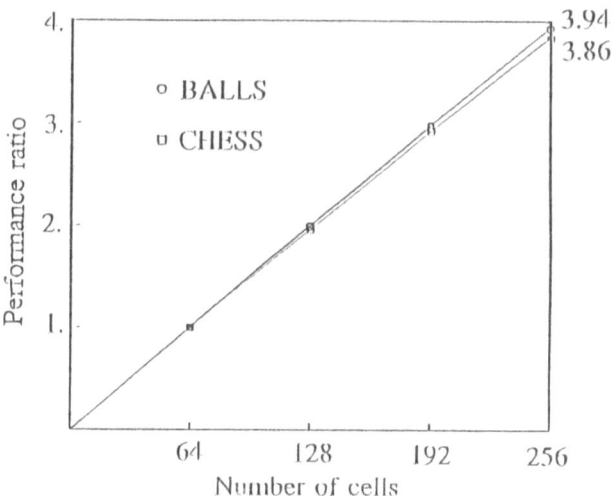

Figure 14 Performance improvement (note that the performance ratio is based on 1.0 when 64 cells are used)

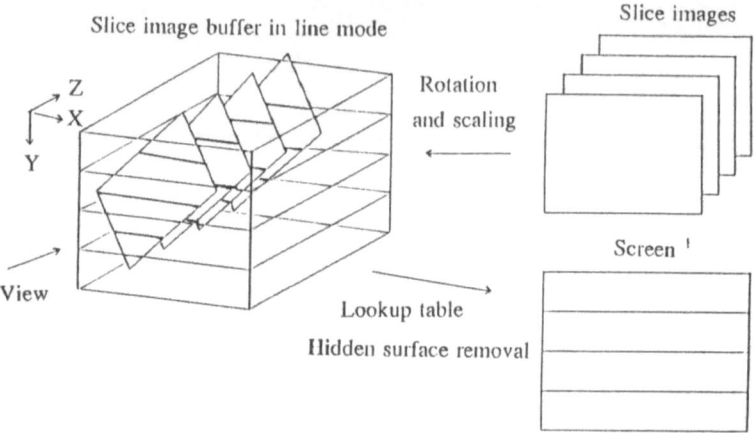

Slice image buffer in line mode

Slice images

Rotation
and scaling

Z
X
Y

View

Lookup table

Hidden surface removal

Screen

Figure 15 Reconstruction of CT slice images

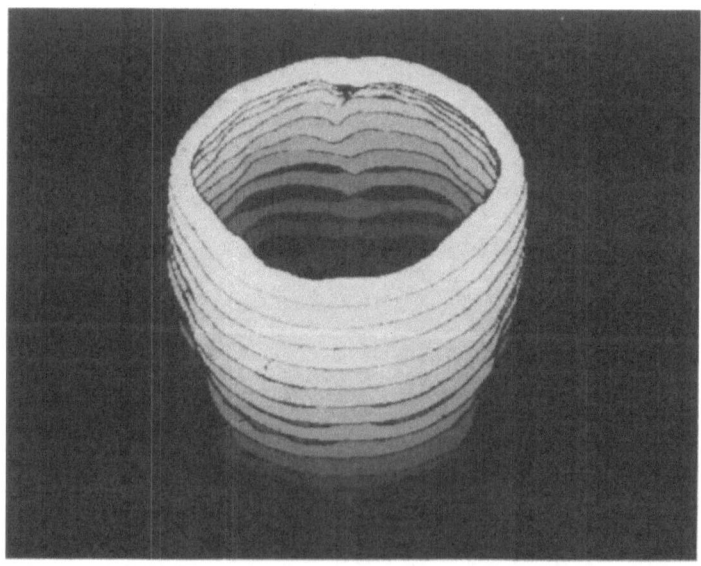

Figure 16 A part of a human skull reconstructed from 12 slices (color art for this figure may be seen in the color insert)

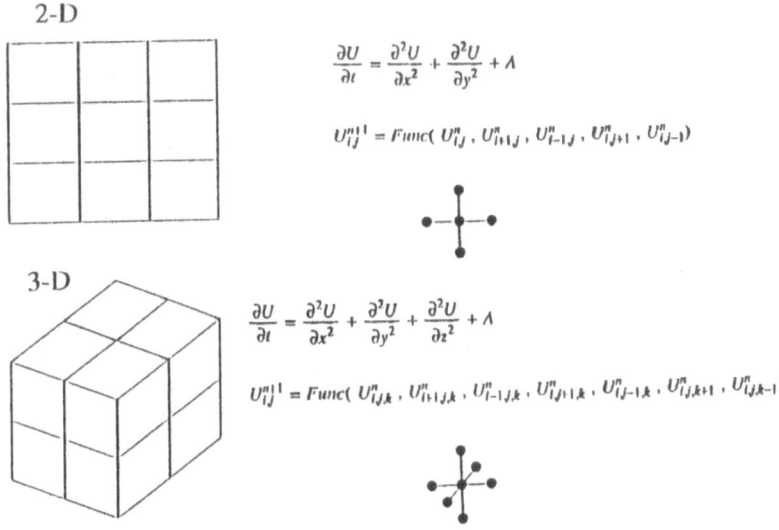

2-D

$$\frac{\partial U}{\partial t} = \frac{\partial^2 U}{\partial x^2} + \frac{\partial^2 U}{\partial y^2} + \Lambda$$

$$U_{i,j}^{n+1} = Func(\ U_{i,j}^n\ ,\ U_{i+1,j}^n\ ,\ U_{i-1,j}^n\ ,\ U_{i,j+1}^n\ ,\ U_{i,j-1}^n)$$

3-D

$$\frac{\partial U}{\partial t} = \frac{\partial^2 U}{\partial x^2} + \frac{\partial^2 U}{\partial y^2} + \frac{\partial^2 U}{\partial z^2} + \Lambda$$

$$U_{i,j}^{n+1} = Func(\ U_{i,j,k}^n\ ,\ U_{i+1,j,k}^n\ ,\ U_{i-1,j,k}^n\ ,\ U_{i,j+1,k}^n\ ,\ U_{i,j-1,k}^n\ ,\ U_{i,j,k+1}^n\ ,\ U_{i,j,k-1}^n)$$

Figure 17 Heat flow simulation

Object space

Image space

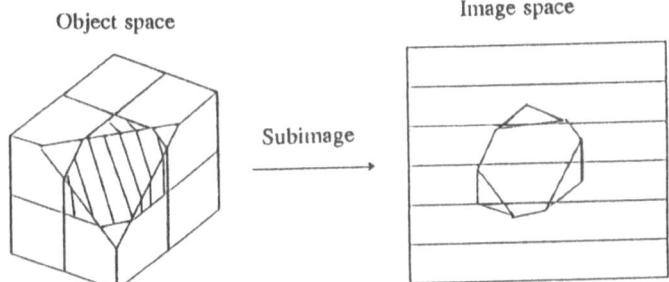

Subimage

Rendering of cross section area

Hidden surface removal in line mode

Figure 18 Parallel processing of cross section display of volume data

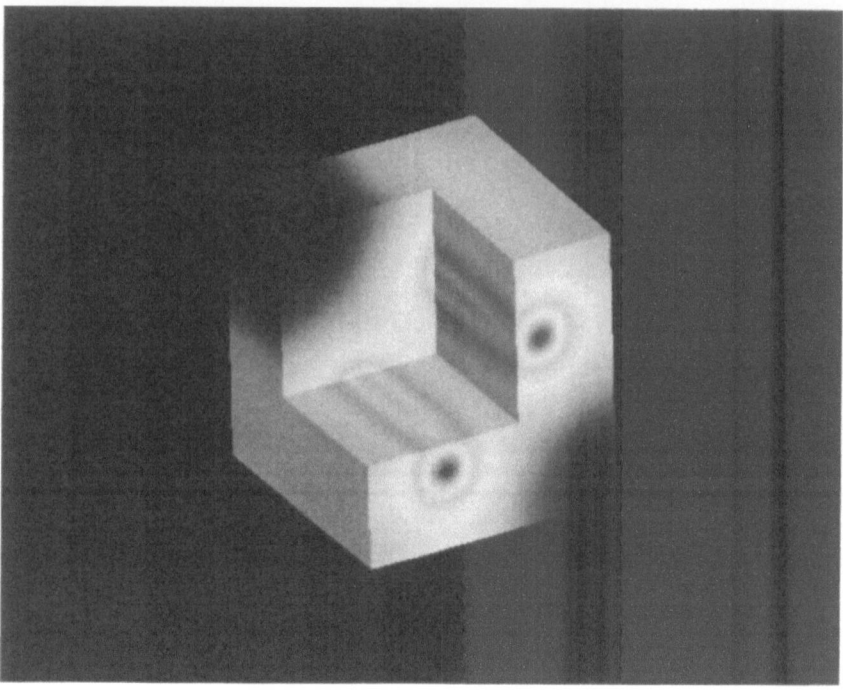

Figure 19 Cross-sectional display of a three-dimensional temperature field (color art for this figure may be seen in the color insert)

Requirements for Scientific Visualization: Evolution of an Accelerator Architecture

Mary C. Whitton
Sun Microsystems, Inc
Research Triangle Park, North Carolina

Scientific visualization has generated much interest in the past year as researchers grapple with the vast amounts of data being generated by supercomputer simulations and data generating devices such as satellites and CT and MRI scanners.   Supercomputers are ill-equipped to help the scientist interactively view data since they seldom contain frame buffers and are frequently shared by many users.

The appropriate computing environment for visualization is a single user, interactive workstation.   Today workstations alone cannot provide the power and functionality required for visualization tasks.   To become a *visualization workstation*, the workstation must be augmented by an accelerator.   This paper looks at the performance and fuctionality requirements for visualization, shows that traditional accelerators do not meet the requirements, and describes an accelerator architecture which does address the requirements.

Visualization is about making pictures of data. It is a new name for a set of well-known and not-so-well-known functions that all involve making images on a computer screen to help man understand and interpret data. The functions that we now collapse into the name *visualization* include *interactive 3-D graphics, image processing, photorealistic rendering*, and the new field of *volume rendering*. In addition, visualization includes techniques such as *stereo display* and *animation playback* which enhance our understanding of 3-D data representations and time sequences of data.

A *visualization workstation* provides the user with basic computing power and functionality, including network connections and local disk storage. Of particular importance for visualization performance are: *fast data paths* to data generation and storage devices; *fast computation* dedicated to preparing data for display and computing pixels; and a *flexible display system* to support the various display formats required by visualization. The performance level of each of these three items is key to an effective visualization workstation.

1.0   Data Path Requirements

1.1   Why connectivity is important

The goal of visualization is to look at data. That data must come to the visualization workstation from somewhere. In addition to data being sourced from local disk storage, it is more and more common for data to be generated and stored remotely. Possible remote sources of data are: supercomputers, network computational servers, 3-D data generating sensors (MRI, CT, PET), and large archival disk storage. Data may also come from a local high bandwidth source - for instance, video rate image acquisition (frame grabbers) or parallel transfer disk.

Fast access to large disks is also needed to retrieve parts of very large data sets interactively, e.g. slices from a 3-D data set or sub-areas of very high resolution maps or photographs. Single data sets can easily run to hundreds of megabytes. Typical is a 6K x 6K satellite image with 8 bits in each of 7 sensor bands for 252 megabytes of data. Data paths need to be bi-directional so that they can support high speed storage of images generated during the visualization cycle.

Data paths internal to the visualization accelerator must be fast and flexible to support the types of access to 2-D and 3-D data sets needed by the variety of visualization algorithms.

1.2   Types and speeds of data paths

The data paths to the visualization workstation need to be high bandwidth. A major goal of visualization is the interactive "steering" of large simulations running on supercomputers. Without high-speed transfer of intermediate results to the visualization workstation for viewing, such steering will be impossible.

The data transfer task in visualization is large. Figure 1.1 shows the effective animation rate for 1024 x 1024 full color images being moved to the frame buffer display at various rates.

Data paths have evolved and data transfer speeds increased with faster bus and network technology. There is better matching now between the transfer task and the speed of the medium. For instance, low speed RS-232 serial lines are seldom used for large data transfers - but remain appropriate for the relatively low speed requirements of human input devices. DMA channels between display device and host offer higher transfer rates, but are frequently bus speed and protocol limited.

Bus Transfer Rate (Megabytes/second)	Effective Animation Rate (Frames/second)
1	1/3
10	3
50	15
100	30

Figure 1.1. Bus transfer rate and effective animation rate

For board level accelerators installed in the chassis of the workstation, a memory mapped interface is possible. Even though the theoretical maximum bandwidth for some buses is quite high (e.g., 40 MB/sec VME bus transfer rate), actual data transfer rates are limited by other factors such as operating system overhead and disk controller speeds.

Extremely high speed channels, such as the HSC with 100 MB/sec, begin to provide the bandwidth necessary for visualization tasks. While very high speed channels may be available on supercomputers, workstations typically do not have the high speed ports required. Development of high speed interfaces is hindered by the lack of an accepted standard.

Fast, flexible internal data paths are also important for visualization accelerators. Data paths must match the requirements of the visualization algorithms. Consider image processing where the data (images) is stored in the display memory. To support filtering operations, the full image must be available in scan line order. Local area access is needed for operations such as adaptive histogram equalization which operate on a small area at a time. Fast random access into the image memory is needed for sophisticated edge-following and image analysis programs.

Networks are essential in today's computing environment, but their speed is inadequate for applications in which large data sets or images must be transferred at rates to support interactivity. Even the new fiber optic FDDI at 100 Mbits/sec cannot meet the demands of visualization.

1.3 System Implications

A visualization workstation must have fast data paths - for external connections to input and output devices and internal connections between processing elements and memory. A hierarchy of connections may be needed: slow for human interaction, fast for connections to supercomputers, local disks, and high speed input devices, and fastest for internal data movement. Storage devices and system software must also support very high bandwidth transfers.

Bandwidth limitations have implications for the development of visualization software. To reduce the amount of data which must be transferred, algorithms must be developed which compress or otherwise minimize data at the source and perform the data expansion as close to the visualization workstation as possible. The data transmission savings can be significant. For example, a molecule described as a series of atom centers and radii requires orders of magnitude less data than a description of the same molecule as solid spheres or as a polygonal surface.

2.0   Computational Requirements

Computations required for visualization are many and varied. The following sections describe some of the computational tasks and give examples of the performance required for interactive visualization. While the requirements are described in MIPS and MFLOPS, neither these metrics nor vectors/second and polygons/second tell the whole story. It is the combination of performance, flexibility, and programmability requirements that make visualization acceleration both unique and difficult.

2.1   Preparing data for image generation

Raw data received from a simulation or a scanner is not necessarily suitable for direct display. Frequently data needs to be processed or massaged before it is ready to be displayed. Significant general purpose processing is required to prepare the data for the image generation algorithm.

For instance, data collected in seismic surveys undergoes lengthy processing to extract the desired information from the original sampled data. The processed seismic data is stored as floating point values which must be converted to integer data in order to compute pixels. Another example is   finite element analysis data computed at irregularly placed nodes being interpolated into regularly spaced voxel data sets

Data expansion is often performed at the visualization workstation. Examples of data expansion are: parametrically defined surfaces expanded into polygonal representation and center-radius representation of atoms being expanded into polygons.

Images are frequently encoded or compressed to reduce storage and transfer bandwidth requirements. The visualization workstation must have the power to decompress and decode images at real time rates to support playback of pre-computed animations.

Floating Point Transformation:	
1 MFLOPs = 1.5 K points/frame	
67 MFLOPs = 100 K points/frame	
Integer Line Drawing:	
1 MIP = 333 ten pixel vectors/frame	
300 MIPs = 100K ten pixel vectors/frame	

Figure 2.1. Interactive 3-D graphics operations for 30 frames/sec update

2.2   Interactive 3-D graphics

Processing requirements for interactive 3-D graphics are well known. Between 8 and 30 frames/second are required for acceptable levels of interactivity. Lines and polygons are accepted standard primitives. Traditional line and polygon graphics is often reduced to hardware - geometry pipeline for transformations and clipping, and drawing engines to accelerate vector and polygon drawing.

Figure 2.1 shows the computational requirements of two fundamental graphics operations: floating point transformations and integer line drawing. Reasonable complex images require significant processing power.

Performance measured in vectors/second and polygons/second keeps going up. But it is not clear that more and more is always going to be better. There is increasing emphasis on the quality of the image drawn rather than only on the quantity of pixels. Anti-aliasing and depth-cueing increase computational complexity, but lead to more easily understood images.

Interactive graphics is no longer limited to just line and polygon primitives. Application specific primitives, such as spheres for molecular modeling and wiggle traces for the seismic market, require that today's graphics accelerator have more flexibility and generality than is available in a hardwired geometry accelerator.

2.3   Image processing

Image processing is dominated by two types of computations: integer operations for real-time manipulation of images and image pairs, and floating point operations such as forward and inverse fast Fourier transforms. Floating point operations have traditionally been performed .on a remote device and hence were not real-time. The advent of fast floating point processor chips has made local floating point capability commonly available.

122

```
Integer Point Operations:

    1 MIP   =  100² area point op/frame
    100 MIPs =  1024² area point op/frame
─────────────────────────────────────────────
Floating Point 3 x 3 Convolution:

    1 MFLOP   =   35² area convolution / frame
    900 MFLOPs =  1024² area convolution / frame
```

Figure 2.2. Image processing operations for 30 frames/sec update

Figure 2.2 gives the computational requirements for two fundamental image processing operations at a 30 frames/second update rate. Very high processing rates are required for interactive rates on large image areas.

Image data can be 8 bit, 12 bit, or 16 bits/pixel. While not needed for all data formats, a 32 bit processor provides the most flexibility.

As with interactive graphics, some image processing algorithms are accepted as standard and can be reduced to hardware. Examples are filtering and ALU type operations on image pairs. More sophisticated algorithms, such as object specific pattern recognition, require increasingly general purpose and powerful computing resources.

2.4   Volume rendering

Volume rendering is a developing and still experimental field. Few "standard" algorithms exist. Three techniques are gaining common acceptance, but much remains to be discovered.

The interior points of a volume can be revealed by making an arbitrary cut through the data. This technique requires very fast access into the volume data set to retrieve the voxels which contribute to the interior slice, and fast trilinear (or more sophisticated) interpolation to compute the intensities at each new pixel.

A second technique displays the surface of a 3-D object as a set of point primitives - a point cloud dense enough to give the appearance of a fully rendered surface. The display step is demanding in its use of transformations and shading based on normals. The preprocessing step of determining normal directions for each point on the surface requires general computational power.

Figure 2.3 shows the computational requirements for tri-linear interpolation and point transformations.

```
┌─────────────────────────────────────────────────────────────────┐
│   Trilinear Interpolation:                                        │
│         1 MFLOPs   = 40 ² area/frame                              │
│         625 MFLOPs = 1024 ² area/frame                            │
├─────────────────────────────────────────────────────────────────┤
│   Point Primitive Transformation:                                 │
│         1 MFLOPs   =      1,500   points/frame                    │
│         67 MFLOPs  =    100,000   points/frame                    │
└─────────────────────────────────────────────────────────────────┘
```

Figure 2.3. Volume rendering operations for 30 frames/sec update.

A third volume rendering technique is to use ray casting and transparency to display multiple interior structures of a volume simultaneously. As with all ray tracing, this technique can consume all available computing cycles.

2.5   Photorealistic rendering

It is arguable whether high quality, photorealistic rendering has a place in general scientific visualization. It clearly does have a place where the science being studied involves lighting, shadows, etc. The techniques of photorealism are valuable in the field of volume rendering.

In general photorealistic rendering is too slow to meet the researcher's need to interact with his data. A lower quality image with higher levels of interactivity is often more valuable.

3.0   Display Flexibility

The display system in a visualization workstation must be flexible enough to meet the needs of graphics, imaging processing, and volume rendering displays. The frame buffer for visualization needs at least 24 planes of color and additional overlay planes for annotation of images. There should be double buffering for smooth motion.

To support stereo viewing, the memory system should be large enough to support quadruple buffering, allowing both the left and right eye images to be double buffered for smooth motion. The control circuitry of the display must allow both field and frame interrupts for buffer switching for stereo.

The system needs look-up tables (LUT's) on each channel of the display to support the image processing requirement of viewing each of the channels as a separate pseudocolor image. In addition, the LUT's provide the capability to perform image enhancement and color correction at the output stage of the display. Many imaging applications involve 12 bit monochrome data and require LUT's with 12 bit inputs.

Video taping of experimental results is a major part of visualization. The display system must support output in television standard formats and be capable of being locked to broadcast quality sync signals.

4.0 Accelerator Architectures

The simplest visualization station is a workstation (or other computer) with a frame buffer. In this organization the CPU does all computations - application processing, graphics transformations, and pixel computation. Pixels are passed to the frame buffer. This is an acceptable configuration for many applications today. Common systems are personal computers with add-on color frame buffer cards. The attractiveness of this configuration is that the CPU has all the generality and flexibility needed for visualization. This simple configuration lacks, however, the power to provide interaction between user and data.

Accelerators are added to workstations to improve performance. Most accelerators are special purpose - dedicated to computer graphics, image processing, or array processing. The goal for each of the special purpose accelerators is to make a limited set of operations more efficient - more nearly interactive. Visualization acceleration needs the capabilities of each of these three special purpose accelerators.

4.1 Graphics Accelerators

Graphics accelerators show the wisdom of moving the computationally intensive portion of the graphics task off the host CPU in that they achieve very high levels of interactivity with very high drawing rates for lines and polygons. Graphics accelerators have the short coming of being limited in functionality. Drawing tasks must be defined in lines and polygons or they cannot be accelerated. Other primitives or more sophisticated drawing algorithms must be executed on the host CPU. Graphics accelerators are generally command driven and are not user programmable. This limits their applicability for visualization.

Data paths in graphics accelerators are often one-way through the pipeline consisting of transformation unit, drawing engine, and frame buffer (Figure 4.1). This makes them unsuitable for algorithms which require reading back data from the frame buffer as.image processing often does.

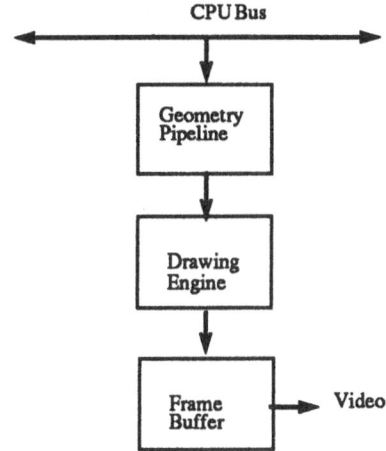

Figure 4.1. Traditional graphics pipeline architecture.

Examples of this type of accelerator are Silicon Graphics, Sun CXP series, and Hewlett Packard graphics workstations.

The new class of "supergraphics workstations" (Ardent, Stellar) is very similar in concept to the basic CPU-frame buffer configuration, yet contains some graphics acceleration hardware. In the supergraphics workstations, the very high performance CPU and floating point units are used both for application processing and for the graphics transformation pipeline. Supergraphics workstations differ from the minimal CPU-frame buffer design by including special purpose graphics pixel drawing hardware in order to support interactive line and polygon drawing.

4.2 Image Processors

Two characteristics dominate the architecture of traditional image processing accelerators: the contents of the image memory banks are available as data and the systems include very high speed integer processing tightly bound to the image memory banks. Multiple memory banks allow the contents of two banks to be used as operands with the results stored back in a third bank. Special hardware for fundamental operations such as convolutions may be included. Figure 4.2 shows a typical architecture.

Image processors are typically supported with a large subroutine library of imaging functions. Operations which are not included in the library must be executed in the CPU and are not accelerated. The fast hardware in the accelerator is not user programmable.

CPU Bus

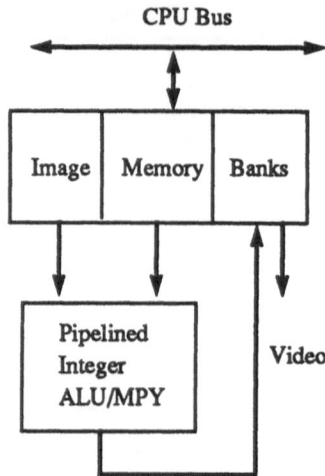

Figure 4.2 Image processor architecture

The data paths between processor and memory banks are almost always limited to accessing the entire image in scan line order. This reduces flexibility for image analysis and interpretation algorithms which require small area and random access into the image data.

Comtal, Vicom, and Gould/DeAnza are examples of traditional image processing products. Androx and VIT offer board level imaging products for engineering workstations.

4.3 Array Processors

Array processor architectures are interesting because of their similarity to Image Processors. The primary differences are that the computation is floating point and there is no display hardware. As in image processing, the data is held in one or two very large memories which are tightly connected to the processing elements. The results are fed back into another portion of the memory. See Figure 4.3.

Array processors are also command driven, most having extensive subroutine libraries of vector arithmetic. High level language compilers for array processors are becoming more common, allowing user programmability and flexible usage.

Examples of board level array processors are products from Sky and Mercury.

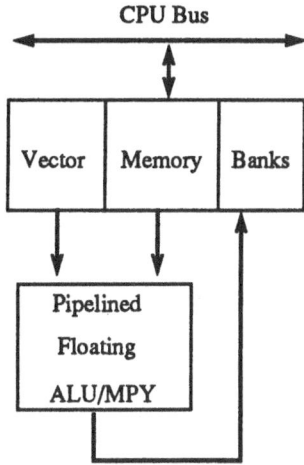

Figure 4.3. Array processor architecture

5.0   A Visualization Accelerator Architecture

A visualization accelerator must contain elements of a graphics accelerator, an image processor, and an array processor.   There are many common elements: the data is frequently a 2-D or 3-D array; high computational speed is needed for integer and floating point operations; and the algorithms are increasing in complexity and in their need for general rather then specialized computation.

Visualization algorithms are just now evolving and cannot yet be committed to inflexible hardware.   Programmability, and preferably user programmability in a high level language, is an important feature.

Visualization accelerator architectures often are very high performance architectures. Frequently the designs incorporate parallel processors or Very Long Instruction Word (VLIW) processors to achieve high levels of performance - and hence high levels of interactivity.   Work continues to develop efficient algorithms for these high performance architectures.

External data paths must include very high speed (100+MB/sec) ports to peripheral devices such as parallel transfer disks and supercomputers. Internal data paths must permit the contents of the memories to be used as data - for both image processing and array processing operations.

The Sun TAAC-1 Application Accelerator and the AT&T Pixel Machine are examples of visualization accelerators.   Pixar, while pioneering in the area of volume rendering and image computing, lacks the essential elements of interactive 3-D graphics and floating point capability.

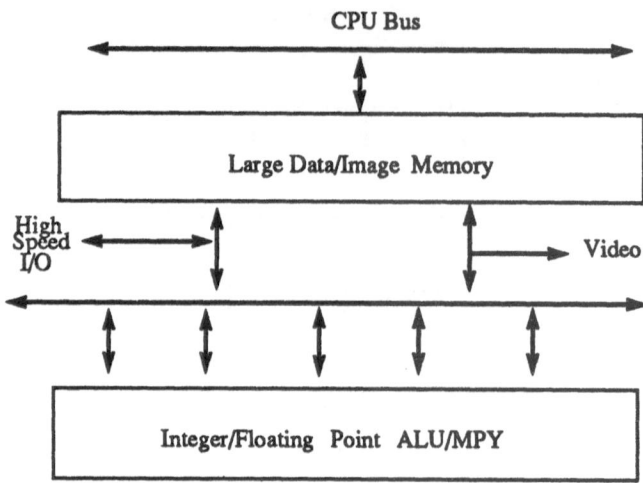

Figure 4.4. Visualization accelerator architecture

6.0 Issues for the Future

Visualization workstations will need to keep increasing in power as the complexity of the algorithms developed for visualization increases. This raises the issue of whether the performance level of uni-processors can keep pace. If the answer is no, then we have real issues of how to perform these tasks in a multi-processor environment. Devices such as the AT&T Pixel Machine are letting us explore such parallel algorithms today.

The role of vendors in the development of visualization hardware and software will be key. The need for collaboration between scientist and vendor is apparent. Only through collaboration will we make significant advances. We vendors don't understand the science thoroughly and scientists don't understand the possibilities in displays as we do. Progress will be made much faster working together - a scientist with a problem working with a graphics/visualization expert with knowledge of hardware and software.

What follows is a true story that illustrates why we need to work together.

When talking with a molecular modeler we asked him about his display requirements. With great conviction he said, "500,000 vectors per second."

Since the machine we currently had to sell didn't meet those requirements, we asked him why he needed so many vectors.

He responded, "To draw lots of detailed contours."

We asked why he needed so many contours and he answered, " I need contours so that I can visualize a surface."

We then asked if he would like to see the surface as a surface - and he said, "Yes, of course, but I didn't think that could be done, so I didn't mention it."

We then asked what he learned from the surface and he replied, "Well, I really want to understand the volume enclosed by the surface and I really know that you can't display volumes directly."

Then we showed him the volume rendering tools and he began to display and understand his electron density data in ways he had never thought of before.

The scientist hadn't described his real problem to us because his thinking was limited to what he believed about display technology. We didn't understand the importance of volumes until he explained it to us.

The bounds will be broken when we work with the scientists as partners to invent new visualization techniques.

Acknowledgments. My thanks go to Nick England and Tim Van Hook for the insights and analysis they have willingly shared over the years we have worked together.

References

1   Akeley, Kurt, and Tom Jermoluk, "High Performance Polygon Rendering." Proceedings of SIGGRAPH88. In Computer Graphics 22, 4 (1988), 239-246.

2   Apgar, Brian, Bret Bersack, and Abraham Mammen. "A Display System for the Stellar Graphics Supercomputer Model GS1000." Proceedings of SIGGRAPH88. In Computer Graphics 22, 4. (1988), 255-262.

3   Chapel Hill Workshop on Volume Visualization, Conference Proceedings (Chapel Hill, NC, May 18-19).   Department of Computer Science, University of North Carolina, Chapel Hill, NC.1989.

4   Deering, Michael, Stephanie Winner, Bic Schediwy, Chris Duffy, and Neil Hunt. "The Triangle Processor and Normal Vector Shader: A VLSI System for High Performance Graphics." Proceedings of SIGGRAPH 88. In Computer Graphics 22,4 (1988). 21-30.

5   Diede, T., C.F. Hagenmaier, G.S. Mirawker, J.J. Rubinstein and W.S. Wozley. "The Titan Graphics Supercomputer Architecture." IEEE Computer 21,9 (Sept 1987). 87-95.

6   England, Nick. "A Graphics System Architecture for Interactive Application-Specific Display Functions." IEEE Computer Graphics and Applications (Jan. 1986). 60-70.

7   England, Nick. "Application Acceleration: Development of the TAAC-1." Sun Technology (Winter 1988). 34-41.

8   England, Nick. "Evolution of High Performance Graphics Systems." To appear in Proceedings of Graphics Interface '89.

9   Fuchs, Henry and John Poulton. "Pixel Planes: A VLSI-Oriented Design for a Raster Graphics Engine." VLSI Design 2,3 (1981). 20-28.

10 Fuchs, Henry, John Poulton, John Eyles, Trey Greer, Jack Golfeather, David Ellsworth, Steve Molnar, Greg Turk, Brice Tebbs, Laura Israel. "A Heterogeneous Multiprocessor Graphics System Using Processor-Enhanced Memories." University of North Carolina Dept. of Computer Science TR 89-005 (Feb. 1989).

11 Fuchs, Henry, et.al. "A Heterogeneous Multiprocessor Graphics System Using Processor-Enhanced Memories." To appear in Proceedings of SIGGRAPH 89.

12   Gharachorloo, Nader, Satish Gupta, Erdem Hokenek, Peruvemba Balasubramanian, Bill Bogholtz, Christian Mathieu, and Christos Zoulas. "Subnanosecond Pixel Rendering with Million Transistor Chips." Proceedings of SIGGRAPH 88. In Computer Graphics 22, 4 (1988). 41-49

13   Levinthal, Adam and Thomas Porter "Chap-A SIMD Graphics Processor." Proceedings of SIGGRAPH 84. In Computer Graphics 18,3 (1984). 77-82.

14   McCormick, B.H., Thomas A. DeFanti, and Maxine D. Brown (eds) "Visualization in Scientific Computing." Computer Graphics 21, 6 (1987).

15   McMillan, Leonard. "Graphics at 820 MFLOPS." ESD: The Electronic Systems Design Magazine. (Sept 1987). 87-95.

16   Potmesil, Michael, and Eric Hoffert. "The Pixel Machine: A Parallel Image Computer." To appear in Proceedings of SIGGRAPH 89.

17   Torborg, John G. "A Parallel Processor Architecture for Graphics Arithmetic Operations." Proceedings of SIGGRAPH 87. In Computer Graphics 21, 4 (1987). 197-204.

18   Upson, Craig and Michael Keeler "V-Buffer: Visible Volume Rendering. Proceedings of SIGGRAPH 88. In Computer Graphics 22, 4 (1988). 59-64.

19   Drebin, Robert A., Loren Carpenter, and Pat Hanrahan. "Volume Rendering." Proceedings of SIGGRAPH 88. In Computer Graphics 22, 4 (1988). 65-74.

20   Van Hook, Tim. "Volume Display Methods." To appear in SIGGRAPH 89 tutorial notes: State of the Art in Data Visualization.

21   Whitton, Mary C. "Accelerating Interactive Applications." Proceedings of NCGA '87. 439-448.

Part 3

Visualization Theory

ADVANCED VISUALIZATION ENVIRONMENTS: KNOWLEDGE-BASED IMAGE MODELING

Bruce H. McCormick
Visualization Laboratory
Department of Computer Science
Texas A&M University
College Station, TX 77843-3112, USA

KEYWORDS / ABSTRACT: scientific visualization / supercomputing / medical imaging / graphics / image analysis / geometric modeling / volume visualizationn / knowledge bases / object-oriented programming / scientific visualization environments

A highly interactive *Visualization Environment* to aid in the visualization, teleconferencing and modeling of massive volumetric data sets is described. The Visualization Environment is designed for applications requiring intensive visualization, image analysis, and geometric modeling. Applications include the interpretation of seismic data in the geosciences, space exploration and astrophysics, molecular modeling, medical imaging, brain mapping, computational fluid dynamics, and microelectronic field modeling.

Three critical aspects of volume visualization and modeling are discussed:

1. The central importance of object-oriented descriptions for visualization and modeling environments;

2. Dynamic finite element modeling of primitive objects in 3D imagery; and

3. Constraint-based assembly of complex models from a set of primitive object models.

The pilot Visualization Environment serves as a testbed for the development and evaluation of these key scientific visualization tools.

135

DESCRIPTION OF AN ADVANCED VISUALIZATION ENVIRONMENT

VISUALIZATION: THE FOREMOST COMMUNICATIONS MEDIUM IN THE WORLD

Visualization -- the foremost communications medium in the world for analyzing and describing scientific, engineering and medical phenomena from the atomic to the anatomic -- is now supported by a computer-based technology. Visualization technology, used in partnership with today's computational tools, whether supercomputer or medical scanner, provides researchers with meaningful information that synthesizes and interprets billions of otherwise unintelligible data values.

We describe here a highly interactive visualization environment to aid in the visualization, teleconferencing, and modeling of massive three-dimensional data sets, which are becoming increasingly common in the computational and imaging aspects of science and technology.

Visualization environments are being designed for applications requiring intensive visualization and geometric modeling, including: the interpretation of seismic data in the geosciences, space exploration and astrophysics, computational fluid dynamics, and microelectronic field modeling. At Texas A&M University we are developing an advanced visualization environment particularly well suited for the visualization and modeling in the biosciences and biotechnology: molecular modeling; X-ray crystallography, microscopy, and medical imaging; and biological structure modeling at subcellular, cellular, tissue, and gross anatomical levels of description.

From the perspective of a user contemplating a spatially oriented visualization and modeling task, the visualization environment provides a natural extension to three and four dimensions of contemporary 2D windowing systems, such as X-windows.

VISUALIZATION OF 3D IMAGERY

In our visualization environment, a "flight simulator" type of interface allows the operator to move about within three-dimensional data sets (surfaces and/or volumes): zooming in for more detail in areas of interest, moving around within an area for a new vantage point, pulling back to establish a new frame of reference, altering display attributes to enhance the information presented, and applying image processing tools to enhance the image further or extract implicit information. This type of environment allows the operator ergonomic control of display attributes and user viewpoint within the data set.

VISUALIZATION OF 2D IMAGERY

As a degenerate case, the visualization environment supports the visualization and geometric modeling of 2D imagery and the interactive networking of visual-based documents and 2D graphics. Additional applications of the visualization environment to the analysis of time-varying imagery are also possible.

KNOWLEDGE-BASED IMAGE MODELING

We have developed a methodology for representing the knowledge associated with complex visual objects in neuroanatomy.[1,2] We propose a similar visual knowledge-based strategy to assist in the interpretation of volumetric data sets that arise in the modeling of other complex structures.

The visual knowledge base can be used as a "program" in the visualization command language to control the accessing of data, windowing, setting of display parameters, and use of image processing tools. The system places under computer control the sequencing of contexts described in the visual knowledge base. The operator is led on the programmed journey and asked to supply his skills at recognizing objects. This recognition task is aided by the following information stored within the visual knowledge base:

* overlay graphics information (2-D and 3-D object models stored in the visual knowledge base and displayed in the approximate position, orientation, and scale);

* text information (characteristic descriptions, exceptions, hints, and advice);

* operator help facilities (menu of cases); and

* automated documentation of operator decisions (especially in unfamiliar data sets).

Accordingly the visual knowledge base provides (1) a cognitive map for navigating through an image data set at multiple levels of resolution, and (2) the knowledge, represented as dynamic models and methods, for the recognition of structures within a volumetric data set.

PORTABLE USER INTERFACE

A workstation-based visualization environment provides an integrated user interface for accessing, manipulating, and modeling three-dimensional image data. The interface is implemented in a Visualization Command Language.[2] Patterned after human visualization, the operations available within the interface allow the user to "visualize" the subject on the screen.

IMAGE DATA MANAGER

A visualization environment serves as a graphical/image data operating system; it supports (1) the processing and modeling of image data, in concert with (2) the design, development, and display of picture files and animation.

The kernel of the visualization environment, the Image Data Manager (IDM), manages and manipulates image/graphical data in the system. The Image Data Manager provides four principal services: data acquisition and management, data analysis, data display, and data transmission. All modules in the visualization environment can request service from the IDM: for network-compatible input/output of image and graphical model data, model creation and editing, image enhancement, data display, and other image/graphical database management functions.

TYPES OF IMAGE DATA SETS SUPPORTED

An advanced visualization environment must support the acquisition of image data sets in different formats from a wide variety of sources and scanning devices, and must store the data sets in suitable formats in its local database. Types of data to be supported include surface/volume graphics models; modeling tools; and experimental data sets: section-based pixel-plane data and volumetric data, as well as 2D graphics, text, and experimental data from other investigators.

This data may then be exchanged with other visualization environment workstations, either by direct data telecommunications or by distribution of downloaded data sets (via optical disks, high density magnetic media, etc.).

We propose now to focus our attention on two of the above attributes of an advanced visualization environment. First, in the next section we introduce the key concepts of *knowledge-based image modeling*. Modeling of volumetric data sets is computationally intensive, and will grow to become a new market for supercomputing.

Second, we look at the *portable user interface*, the babel of languages/systems for graphics and imaging. We ask how we might bring order out of chaos -- without hypothesizing herculean efforts at standardizations.

Finally, in the last section we summarize the *advanced features of a visualization environment*.

KNOWLEDGE-BASED IMAGE MODELING

The problem of reconstructing the three-dimensional structure of an complex object is pervasive in many areas of visualization and computer vision research.

Several imaging technologies, such as medical imaging and laser scanning confocal microscopy, naturally provide volumetric data sets. Physical fields are intrinsically three- (or four-) dimensional, and their numerical integration by a supercomputer yields comparable volumetric data sets. Considered volumetrically, a complex object such as the brain (at $(40 \text{ um})^3$ per voxel resolution) and a fluid flow field are comparable data sets; however, the former has no known differential expression.

FROM PDE MODEL TO VOLUMETRIC DATA SET

Contemporary scientific visualization starts with a set of partial differential equations (PDEs) and boundary conditions typically described over mesh-defined surfaces. For example, in computational fluid dynamics the key equations are the Navier-Stokes equations, and the surfaces may be those of the space shuttle or a fighter aircraft. The boundary mesh is then extrapolated into a 3-dimensional cellular array and the finite element method is evoked to solve the PDEs over the cellular array.[3,4] For example a scalar field may be computed at the nodes of the finite elements. From the 3D cellular representation, then, we can interpolate from the nodes to compute the appropriate value of the scalar field at each voxel of the image data set. In summary, we have transformed a continuum PDE model into a discrete, digitized volumetric data set:

PDE model --> finite element nodal values -->voxel values.

In this forward modeling scheme, surface elements corresponding to high gradients of the scalar field can be computed directly on the cellular model, and the integrated surface built up in this representation. That is, the voxel-based representation need play no direct role in the computation.

FROM VOLUMETRIC DATA SET TO IMAGE MODEL

In imaging, we face the inverse situation. We know the scalar, vector, or tensor field at every voxel of the volumetric data set, and we must now infer a model which could have given rise to this volumetric data set.

In general, detail is not of uniform granularity throughout the volumetric data set. For example, in anatomical data sets (such as stained histological sections of brain) one can see a natural coordinate system, which partitions the sectional data into cells conforming to the natural layering of the tissue (that we see, for example, in the 6 layers of the cerebral cortex, or the 10 layers of the retina).[5]

But these natural coordinate systems are only locally defined within each object (for example, each thalamic nucleus has its own globular boundary).

We propose to develop a visualization environment to facilitate the editing and modeling of 3-D volumetric data. These models must be deformable so that the model can be deformed dynamically to match the image data set. A major difficulty in solving this image modeling problem is the insufficiency and ambiguity of visual information in any given image data set. Therefore we must aim to maximize our use of what is, after all, actually abundant visual information provided by the 3-D volumetric data set and our prior visual knowledge of the domain. The specific aims of the proposed work include the following tasks:

1. *Model-independent Representation of Complex Objects:* Knowledge-based modeling uses hierarchical, object-oriented representations.

2. *Dynamic Model Construction:* Deformable 3-D models represent object primitives, which can be deformed for dynamic matching.

3. *Dynamic Matching:* Pseudo-fields deform models for matching with volumetric image data.

MODEL-INDEPENDENT REPRESENTATION OF COMPLEX OBJECTS

How is it that we teach students to form a cognitive model of complex objects? How do we communicate the knowledge associated with a visual subject? To study this question, we focused our attention on how illustrated textbooks teach anatomy and neuroanatomy. One text in particular, K. E. Moyer's textbook, *Neuroanatomy*, was selected for this research. The text provides a valuable source of the language and figures that communicate the highly visual subject of human neuroanatomy.

Neuroanatomy is organized around sections, with each section concentrating on one or two regions of the human nervous system. A section typically consists of two facing pages: text on the left and a figure on the right. Because of the visual nature of neuroanatomy, the lesson presented in the text is organized around the labeled objects of the figure.

We note what is presented by the text, and equally important, what is not. Objects of the nervous system are presented in hierarchical order. The taxonomy of the brain parts forms a lattice (not a tree), so the text uses a hierarchical, object-oriented representation.

What is not presented in the text is any graphical model of the floppy, or deformable, brain parts. For our purposes, they could be modelled by finite element meshes, generalized cylinders, superquardics, and possibly other approximate representations.[6,7,8,9] But none of this is relevant to the text, because in this first stage of modeling, which emphasizes the taxonomy and spatial relationships of the parts of the complex object (in this case, the brain), no explicit representation is needed other than the illustration itself. The representation at this stage is model independent.

Koons, in his 1988 thesis,[2] summarizes the results of his study of visual knowledge acquisition from illustrated texts, as follows:

"Three hypotheses, formed during the analysis of Moyer's textbook *Neuroanatomy*, provide the foundation for a representational model of visually-oriented knowledge. First, the human visual system is limited in the amount and complexity of the information that it can hold and process at any one time. When confronted with a complex subject, the visual system divides the scene into a set of simpler views. The information within each of these windows can now be processed, resulting in an abstracted concept that captures the theme and character of the window's content. Second, the transition between two concepts can be defined by the changes in the focus of attention. By recording the changes in the location and size of the attentional window, the spatial and logical relationship between the concepts can be represented. Third, in order to record the context of each concept in a complex scene, the mind maintains a structure or graph of concepts based on the observed organization, thus allowing the representation of the knowledge for structuring or organizing a visual subject such as neuroanatomy.

The representational model for visual knowledge is based on this two-part definition:

Visual Frames: A frame captures a visual concept by encoding the relevant visual attributes from within the current attentional window, and the associated meaning or interpretation for that information.

Attentional Links: Links connect visual frames to capture the perceived or learned organization of a complex visual subject. A link between two frames is recorded as the operations (PAN and ZOOM) for changing the location and size of the attentional focus.

Within a frame, the associated visual concept is represented in a floppy model. Similar to the two-part model of visual knowledge, a floppy model is represented as a limited number of spatially distributed samples linked together into a structure. The placement of samples within the attentional window concentrates on those areas that uniquely characterize the subject. A link, connecting two samples or nodes in the model, is represented as a constraint on the relative movement of the two nodes. By specifying the strength and allowable range of each constraint, a flexible or deformable model of the subject can be constructed that will describe an object or shape and its acceptable variations. The model also allows for the future use of information from morphometrics, measured properties of tissues and materials, and biological coordinate systems to produce highly accurate dynamic models of anatomical subjects."

DYNAMIC MODEL CONSTRUCTION OF OBJECT PRIMITIVES

Variance across samples of an object is inevitable and limits the expressibility and generality of a rigid modeling scheme. To cope with the inevitable variances, a model must be flexible in such a way that dynamic deformation can be applied to it when the model is matched with the image data. [10,11,12,13,14, 15,16,17]

What constitutes a good model? For anatomical data one knows, for example, that the thalamus will vary significantly in shape from one specimen to another. A good model should carry a natural coordinate system, adapted independently to each specimen, such that landmarks (if any) correspond, stain boundaries correspond, and in general that the tissues at identical coordinates in the two coordinate frames are as biologically homologous as possible. For example, for drug studies the distribution of drug-sensitive sites in the tissue should correspond. This done, we are ready to enter the age of biological CAD.

In general we require of models of object primitives the following properties:[5]

* Spatial relations to neighboring objects are preserved, subject to the appropriate constraints.

* In their respective coordinate systems, landmarks correspond.

* If deformation occurs, then the physical model should be consistent with the physical processes (fluid flow, elastic deformation) or biological processes (growth) that account for the deformation. That is, the object models should be covariant with the objects under physical/biological deformation.

In summary, we would like:

volumetric data set --> models providing natural coordinate systems --> description of local physical/biological processes.

This modeling activity is quite as intensive computationally as the forward computation of the field over the finite element mesh. There seems little likelihood that such descriptions can be built ad hoc without prior knowledge of the domain, except for very simplified examples such as the separation of bone from soft tissue in CT scans. We spend enormous energies to teach medical students how to identify brain structures in multiple specimens of the brain -- to provide a cognitive map of the brain -- which would be unnecessary if these brain parts could be reliably identified by human or computer without this visual knowledge base.

DYNAMIC MATCHING

Although there is much visual information in a volumetric data set, it is difficult and inefficient to manage such large amounts of visual data in a single-stage interpretation process. Dividing a 3-D image into thick slices seems to be a reasonable compromise: the amount of image data is manageable, while better contrast and less ambiguity are retained than if individual 2-D sections are sequentially examined. The image can be matched thick slice by thick slice, and a 3-D volumetric model constructed from the series of matched thick sections.

Typically, layers characterize the sectional data of a complex object such as a brain or biological cell cluster.. Each of these surface layers can be represented as a rubber band. The elasticity of the "rubber band" can be prescribed to permit dynamic matching with the sectional image data. A three-dimensional model of the thick section of tissue can then be constructed from the nested surfaces, or layers, described by these rubber bands. In this way, the model is capable of describing complex objects in sufficient detail and yet remains manageable in terms of its computational complexity and memory requirements.

By incorporating the theory of elasticity into representation of the object, we can achieve the desired deformability of the model. The elasticity of each rubber band can be dynamically adjusted after comparing the fit of the model to a set of sample objects of the same class. The allowable range of variation for each band can then be controlled through the parameters of the elasticity of the object model.

To synthesize an image, the matched sectional models must then be visualized in three-dimensional volumetric form. Therefore the visualization environment must be capable of rendering volumetric data from thick slices of the matched 3-D model into an image that allows interactive three-dimensional viewing of the volume. This can be accomplished with volume-rendering software, such as currently available on the Pixar Image Computer.

For a full 3-D image, the goal is more ambitious than a simple surface computed by the modeling activity, even though this may be adequate for modeling a joint for a hip replacement operation. In general one wants to cover the data set by object primitives, each having its own natural coordinate system.

EVOLUTION TOWARD AUTOMATED IMAGE MODELING

Moving portions of the interpretation task under machine control allows the following benefits:

* The researcher faced with the interpretation of volumetric imagery is provided with an intelligent and responsive visualization environment. This environment relieves the researcher of tedious time-consuming image analysis and modeling tasks and greatly enhances productivity.

* Any human decision involves a degree of subjectivity. By exactly and clearly documenting the decisions involved, we achieve a better understanding of the task. By automating these decisions, we will achieve a higher degree of consistency. Exchange of data sets, models and results among the worldwide scientific community is thereby facilitated.

* Finally, the expert image interpreter's time is expensive and very limited. By capturing this knowledge, we can replicate the required expertise as many times as needed to achieve the desired production of image analysis and modeling.

In summary, the visual knowledge base provides (1) a cognitive map for navigating through a complex volumetric data sets at different scales of resolution, and (2) the dynamic models and methods for the recognition of particular objects within the volumetric data set. Early phases of this work will include human interaction to check the machine's decisions and to evaluate the performance for improvement or alterations. Later versions of the visualization environment will shift image identification work more and more toward machine control, reducing operator involvement, minimizing human subjectivity, and increasing the productivity of the system.

PORTABLE USER INTERFACE

BABEL OF VISUALIZATION LANGUAGES

There are many -- perhaps too many -- languages/systems for graphics and imaging. For example, we have: PostScript for 2D text and figures; PHIGS (and PHIGS+) for 3D graphics; X-windows (version 11) for windowing: the IRIS 3D surface graphics library; and the PIXAR 3D surface/volume rendering and image processing libraries. One can easily compile a list of 30 languages or command libraries, each of several hundred commands, that are all reasonable candidates for incorporation in a visualization environment. Faced with this babel of visualization languages, what strategy should we adopt to develop an optimal visualization environment?

Several possibilities come to mind:

1. Standardization by government or international organization will eventually enforce one standard; or perhaps,

2. One firm will gain sufficient market share to impose its view of visualization on the world; or, in the shorter run,

3. A market economy for visualization software will arise much like that for word processing, spreadsheets, and desktop publishing in the personal computer industry.

This third option is, I believe, the most likely for the next decade. Even in the personal computer marketplace, integrated software (such as Symphony, Jazz, and Microsoft Works) has failed to dominate because (1) integrating multiple subsystems is much slower work than developing independent modules, and (2) the non-integrated modules have so rapidly improved in functionality as to prevent their ready incorporation in an integrated environment. In practice, integration imposes mediocre performance upon all components.

AGREEMENT ON A COMMON INTERFACE

If we are unlikely to have a standardized complement of visualization command languages, how can we avoid overburdening the user?

1. Commercial visualization languages/systems can be generalized by an *object-oriented language/ system description*. By introducing object-oriented descriptions, we can compress these languages from 200-300 commands down to a matrix of 10-15 generic commands applied to 10-15 object classes.

2. We can agree on a *common interface* to evoke any of the visualization subsystems (paralleling the role played by the "look and feel" of the Macintosh). From the common *object-oriented* interface, all generic commands -- regardless of language/system -- can then be evoked, and

using hypertext techniques, appropriate documentation can be embedded in accompanying windows.

3. We can view the visualization interface as a *software bus* into which an assortment of commercial visualization subsystems can be plugged. All modules would be evoked from the common interface operating through your favorite windowing system.

Our experience with object-oriented description of visualization languages/systems is still quite limited. We have experimented with three languages: PostScript, X-windows, and the PIXAR image computer library of commands. All have taken approximately two man-months to map from the 200-300 commands to the generic function/class object matrix format. Though this work is still in an early stage, we believe that this mapping can be largely automated.

COUPLING TO THE IMAGE DATA MANAGER

The kernel of the visualization environment, the Image Data Manager (IDM, see above), manages and manipulates image/graphical data in the system. Many real difficulties of establishing a common software bus for visualization environment modules center on developing a common data interface to the IDM, supplied by the workstation vendor. In our view workstation vendors will increasingly be forced to standardize this data interface as they try to optimize local and global networking.

PostScript has offered us a first glimpse of a structured image data description language. Because of its dominance of the laser printing market, PostScript has become a de facto standard. Its implementation on the laser printer is considered a vendor responsibility. The user assembles visualization modules that speak "PostScript"; their subsequent translation into printer output or monitor screens is not the user's concern.

Generalizing this methodology, we should insist that all our visualization modules address image data through a generalized *image data description language* (in other words, a three-dimensional PostScript). Our core visualization workstation would provide a hardware-dependent IDM, but one that would understand the generalized image data description language. We could then acquire our choice of visualization modules from a marketplace common to all visualization workstations, regardless of vendor.

PostScript has introduced a layering which isolates 2D image data description from its printing and display. We must now generalize this layering to 3- and 4-dimensional image data description. In particular, we must isolate 3D surface/volume rendering from the idiosyncrasies of hardware implementation. The vehicle to impose this separation is the generalized image data description language.

ADVANCED FEATURES OF VISUALIZATION ENVIRONMENTS

DO NOT CONFUSE VISUALIZATION WITH MODELING: MODELING IS MUCH HARDER

Visualization informs the eye, but does nothing to provide a parametric description or summary of the relevant objects in the data set, of the kind made possible by geometric or dynamic modeling. Visualization constitutes the necessary first step toward image modeling. A visualization environment which in addition supports image modeling is considered "advanced."

VISUALIZATION AND MODELING OF VOLUMETRIC DATA SETS ARE COMPUTATIONALLY INTENSIVE PROCESSES

A 2-dimensional visual image of 512 x 512 pixels x 8 bits/pixel contains approximately 100 times more information than a screen of text with 25 rows x 80 characters/row. Volumetric images, approaching 1024 x 1024 x 1024 x 4 bytes/voxel, present approximately 16,000 times more information than a single 512 pixel image. In addition dynamic modeling uses relaxation techniques, which are computationally intensive.

SUPERCOMPUTERS PROVIDE A GOOD MODEL OF THE COMPUTER ARCHITECTURE FOR ADVANCED VISUALIZATION ENVIRONMENTS, AND A POOR MODEL FOR THE SOFTWARE

The Sparc chip architecture of the Sun 4 Unix workstation, has been described by Bill Joy, Vice President for R&D at Sun Microsystems, as modeled on the RISC instruction set common to supercomputer architecture. Register arrangements, cache organization and compiling techniques learned in the supercomputer arena are all being transferred to advanced graphics workstations. Meanwhile, high performance workstations continue to trail supercomputer performance by approximately an order of magnitude.

Advanced visualization environments, however, call for a highly interactive user interface and an extensive software base that will be most profitably developed for the extensive advanced personal computer/workstation market.

KNOWLEDGE-BASED IMAGE MODELING EXPLOITS HIERARCHICAL, OBJECT-ORIENTED REPRESENTATIONS

Twenty- five years of bottom-up image analysis, starting from low-level image processing, failed to produce acceptable results in computer vision. In the past two or three years, however, object-oriented dynamic modeling of 2D imagery has produced several successes. Extension of these techniques to hierarchical object-oriented dynamic modeling of volumetric data sets looks very promising.

DYNAMIC MATCHING USES PSEUDO-FIELDS TO DEFORM THE MODEL INTO THE CORRECT SHAPE

Edge fragments and isolated surface normals in the volumetric data set give evidence of the existence of underlying objects. These primitive elements are assembled to generate a pseudo-field, which in turn deforms the dynamic object model to assume its energy minimum -- that is, its best representation of the object. In summary:

Volumetric data set --> pseudo-field --> object model.

DOMAIN-SPECIFIC KNOWLEDGE CAN BE USED TO GUIDE AND AUTOMATE THE DYNAMIC MATCHING PROCESS

The allowable deformations of an object, exhibited over an ensemble of sample data sets, can be captured in the relative elasticity of links in a finite element model of the object. With more examples, a more appropriate model can be learned, and a more discriminating match can be made of the model to the object data set.

ADVANCED SCIENTIFIC VISUALIZATION REQUIRES A GENERIC IMAGE DATA DESCRIPTION LANGUAGE

A single standardized Visualization Environment is highly unlikely to dominate the market in the next decade. More reasonably, we should seek to standardize on a "software bus" into which can be plugged visualization modules supplied by a variety of vendors to give the desired level of functionality.

First steps toward this evolution are to give current visualization subsystems an object-oriented description and provide a common user interface to these object-oriented descriptions.

A later step is to layer image data management in the visualization environment by providing a standard generalized image data description language used by all visualization modules.

REFERENCES

1. Koons, D. B. and B. H. McCormick, A model of visual knowledge representation. *Proc. First International Conf. on Computer Vision,* 1987, pp.365-373

2. Koons, D. B., *A model for the representation and extraction of visual knowledge from illustrated texts.* MS Thesis, Report TAMU 88-010, Department of Computer Science, Texas A&M University, August, 1988.

3. Burnett, D. S., *Finite element analysis.* MA: Addison-Wesley 1987.

4. Zienkiewicz, O. C., *The finite element method (3rd edition).* Maidenhead, England: McGraw-Hill, 1986.

5. Tiwary, A., *Understanding biological form: A knowledge based environment to study form change.* MS Thesis, Department of Computer Science, Texas A&M University, December, 1986.

6. Mortenson, M., *Geometric modeling.* NY: John Wiley & Sons, 1985.

7. Bartels, R., J. Beatty, and B. Barsky, *An introduction to splines for use in computer graphics & geometric modeling.* CA: Morgan Kaufmann, 1987.

8. Barr, A., Superquardics and angle-preserving transformations. *IEEE Computer Graphics and Applications,* Jan., 1981, pp. 11-23.

9. Pentland, A., Perceptual organization and the representation of natural form. *Artificial Intelligence* 28, 1986, pp. 293-331.

10. Terzopoulos, D., A. Witkin and M. Kass, Constraints on deformable models: recovering 3D shape and nonrigid motion. *Artificial Intelligence* 36, 1988, pp. 91-123.

11. Terzopoulos, D., J. Platt, A. Barr and K. Fleisher, Elastically deformable models. *Computer Graphics,* July 1987, pp. 205-214.

12. Terzopoulos, D., A. Witkin and M. Kass, Symmetry-seeking models for 3D object recognition. *Proc. IEEE 1st Intl. Conf. on Computer Vision,* 1987, pp 269-276.

13. Terzopolous, D., Regularization of inverse visual problems involving discontinuities. *PAMI* July 1986, pp 413-424.

14. Terzopoulos, D., Multilevel reconstruction of visual surfaces: variational principles and finite element representations. MIT A.I. Laboratory, AI Memo 671, 1982; Reprinted in *Multiresolution image processing and analysis,* A. Rosenfeld, Ed. NY: Springer-Verlag, 1984, pp. 237-310.

15. Witkin, A., K. Fleisher and A. Barr, Energy constraints on parametrized models. *Computer Graphics,* Vol. 21, No. 4, July 1987, pp. 225-232.

16. Barr, A., Global and local deformations of solid primitives. *Computer Graphics*, Vol. 18, No. 3, July, 1984, pp. 21-30.

17. Sommerfeld, A., *Mechanics of deformable bodies.* NY: Academic Presss, 1950.

GEOMETRY VS IMAGING:
EXTENDED ABSTRACT

Dr. Alvy Ray Smith
Cofounder and Executive Vice President

Pixar

GEOMETRY VS IMAGING†

There are two quite distinct ways of making pictures with computers. The geometric way is quite widely understood - and often thought to be the only way. The imaging way is less intuitive - and leads to a different marketplace which is probably as large or larger than that for geometry. The terminologies, theories, and even heroes of the two worlds are quite distinct, and the hardware devices to implement them are strikingly different.

Figure 1 illustrates the two domains and their interrelationships. Geometry-based picturing begins with the description of objects or scenes in terms of common geometric ideas - polygons, lines, spheres, cylinders, patches, splines, etc. Recall that these are mathematical abstractions, not pictures. To make a digital picture of a geometrically described object requires that it be *rendered* (or *rasterized* or *scan converted*) into pixels. Geometric concepts live in real continuous space, requiring floating point for accurate computer representation. Famous names are Pythagoras and Euclid. The theorems of analytic geometry are of paramount importance.

Imaging-based picturing begins with a set of discrete samples - pixels - of a continuum, usually placed on a uniform grid. As Figure 1 shows, these samples *may* come from scan conversion of geometry, but in general they do not. In the majority of cases they come from non-geometric sources such as digitized satellite photographs, computed tomographic (CT) or magnetic resonance (MR) medical scans, digitized X-radiographs, electronic paint programs, seismic sensors, supercomputer simulations of partial differential equation systems, or laboratory measurements. In all cases, an array of numbers (samples) is the original data - not a "display list" of geometric primitives. Imaging generates pictures from this data by directly displaying them on the computer screen. The imaging domain is discrete by definition

---

† The full text of this paper can be found in *Computer Graphics World*, November, 1988. Reprinted with permission.

and integer arithmetic typically suffices. Famous names are Nyquist and Fourier. The Sampling Theorem is of paramount importance.

As Figure 1 also points out, it is possible in some cases to extract geometric data from sampled data and reenter the geometric domain (and then render the geometry to reenter the image domain!) This step is not required, however, to make pictures - contrary to the opinion of a surprisingly large number of people. In fact, it frequently introduces thresholding artifacts (jaggies) which may be highly undesirable - as in medical diagnostic imaging which insists, and depends, on no alterations of its data. Direct imaging of data arrays avoids such artifacts.

Notice that the distinction between geometry and imaging is *not* that between image synthesis and image analysis. An electronic paint program is an excellent example of a non-geometric synthesis technique - an imaging technique. Nor is it the distinction between "computer graphics" and "image processing". Image processing is only a subset of imaging, and computer graphics loosely covers picturing from both domains. The fundamental difference is whether the elemental datum is a geometric or a numeric entity - a polygon or a pixel.

ARCHITECTURAL DIFFERENCES

The geometry and imaging distinction is reflected in special-purpose hardware *accelerators* available for each. All graphics computations discussed here *could* be implemented on a general-purpose computer, such as a workstation host, minicomputer, mainframe, or personal computer. But in the late 1980s, it is still the case that these offer insufficient price/performance for geometry and imaging computations. The general-purpose machines with relatively lower computational power simply cannot do the computations in a tolerable amount of time or lack sufficient memory; the more powerful machines use cycles which are too expensive compared to what can be purchased for much less in accelerators.

Geometry accelerators are measured in terms of the number of geometrical objects they can manipulate in realtime. Imaging datasets are typically so large that realtime is not yet an appropriate measure for them (unless programmability is sacrificed). So imaging accelerators are measured in terms of the number of pixels they can comfortably manipulate on the order of 100 times faster than a host computer. Some geometry engines do a little imaging, and some imaging computers do a little geometry. The purpose of the next section is to clarify the distinctions.

SOPHISTICATION METERS

Figures 2 and 3 are *sophistication meters* for geometry and imaging respectively. They are attempts to summarize the major techniques and terms in the two domains. Both charts are ordered from bottom to top in each column by increasing sophistication. The horizontal lines in the columns mark the *realtime lines* for each chart. These are placed somewhat generously and are explained more fully below. Notice that the terms used on the two charts are almost completely different.

Geometry Sophistication Meter

The capabilities of realtime geometry accelerators in 1988 lie below the realtime lines of Figure 2. The most obvious observation is that most of what is known about geometry-based graphics has not yet been pulled into realtime. This is particularly true of the "shading" of geometric objects - their visual content - which is in general more difficult than the shaping of the objects - their geometry. The recently proposed RenderMan Interface is exactly a roadmap to all the non-realtime geometry-based graphics.

Imaging Sophistication Meter

The first thing to notice about Figure 3 is that sophisticated imaging applications are tremendously complex in terms of pixel count. This is the world addressed by imaging computers. Some geometry accelerators confusingly offer restricted imaging capabilities as well as geometry. The realtime lines in this chart indicate where in the scheme of things this limited imaging lies. In particular, it is restricted to essentially 1.25K×1K display memories. Of course, the host computer in geometry workstations can always do the imaging - by definition of computing - but then it executes at general-purpose price/performance, not the order-of-magnitude lower price/performance of a special-purpose imaging accelerator. This is standard "you get what you pay for" general-purpose computing.

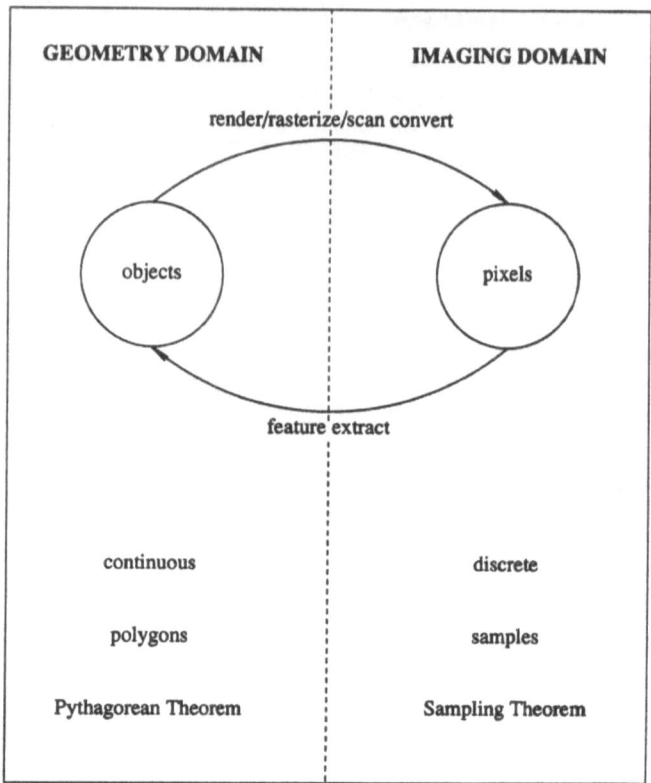

Figure 1

GEOMETRY SOPHISTICATION METER

Shading "Shade"	Geometry "Shape"	Anti- aliasing	Complexity #Primitives Per Picture
shading language			
materials			
shaped lights			
distributed lights			
matte/glossy			
displacement maps			
environment maps			
bump maps			
image texture maps			
radiosity			
refraction			
procedural textures			
Phong shading	hyperpatches	motion	100,000,000
transparency	patches	textures	10,000,000
Gouraud shading	quadrics	specular	1,000,000
multiple lights	nurbs	edges	100,000
flat shading	polygons	lines	10,000
none	lines	none	1,000

Used with permission of Pixar

Geometry sophistication increases up each column, from bottom to top.

The state of realtime geometry machines in 1988 is below the lines in each column. The RenderMan Interface addresses the entire chart with particular emphasis above the lines.

Figure 2

IMAGING SOPHISTICATION METER

Techniques	Dimensions	Filtering	Complexity #Pixels Per Image
3D painting, FFT, and compositing			
volumetrics			
volume imaging			
warping			
classification: thematic, MR, and CT			
FFT, Walsh, and other transforms			
compression			
soft-edged painting and soft fill		very wide	
convolutions and filters		bessel	1,000,000,000
		sinc	100,000,000
histograms and	volume movies	cubic	10,000,000
equalization	volumes	gauss	1,000,000
point operations	image movies	box	100,000
matte algebra	images	none	10,000

Used with permission of Pixar

Imaging sophistication increases up each column, from bottom to top.

This chart measures imaging accelerators, but some *geometry* accelerators do limited imaging - below the lines in each column. Image computers can address the entire chart.

Figure 3

Lighting Simulation

Eihachiro Nakamae
Hiroshima University

1 Introduction

Beginning with the publication of Ivan Sutherland's interactive graphics system "Sketch-pad" in the early 1960's, researchers began to attack the problem of generating real-istic computer graphic images. The fundamental problem of hidden line elimination was solved by the end of the 1960's. Beginning in the early 1970's, advances in hardware development, particularly raster scan CRT's with integrated frame buffers, provided researchers with excellent tools. Since that time a wealth of techniques for generating realistic images has been developed; hidden surface elimination, 2D and 3D texture mapping, shading, shadowing, and modeling of natural objects such as trees, clouds, and mountains.

The principle problems to be dealt with in computing shading and shadows are the simulation of various light sources, mutual inter-reflection (radiosity), haze, and visible shafts of light. In this paper we will discuss calculation of lighting and shadow effects, and application of parallel processing techniques.

2 Shading [1],[2]

Illuminance consists of two components, direct and indirect. Illuminance must be calculated for both shadowed and unshadowed regions.

2.1 Illuminance Calculation in Unshadowed Region

a) Point Light Sources

The luminous intensity of a point light source can be totally characterized by a luminous intensity distribution curve. This curve represents the variation of luminous intensity of the source over a plane passing through the center of the source.

The diffuse illuminance at an arbitrary point on an object's surface due to a light source is a function of the luminous intensity of the light source, the distance from the light source, and the angle between the surface normal at the calculation point on the surface and the light ray (See Fig. 1 (a)).

b) Linear Light Sources

A linear light source can be considered to be composed of an infinite number of point light sources lying on a line segment. Therefore, the diffuse illuminance at an arbitrary point on an object's surface due to a linear light source is obtained by integrating

157

158

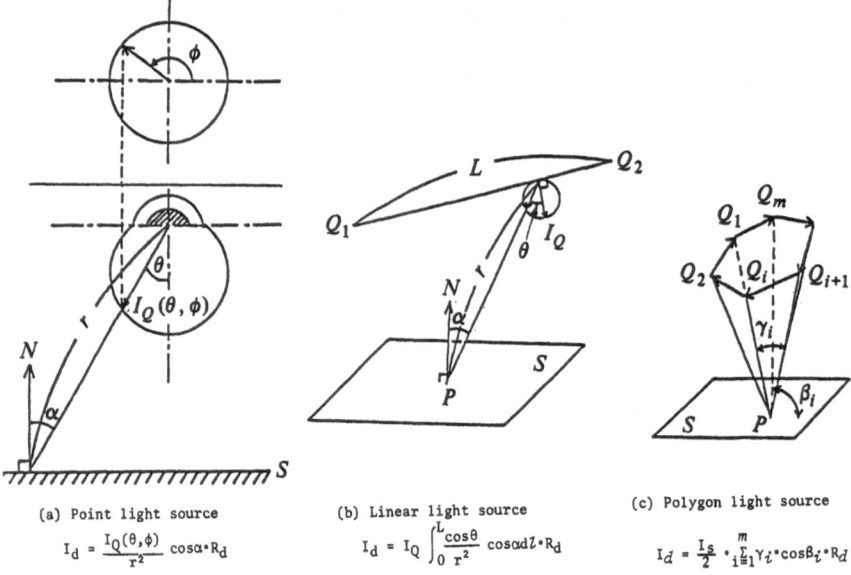

(a) Point light source

$$I_d = \frac{I_Q(\theta,\phi)}{r^2}\, \cos\alpha \cdot R_d$$

(b) Linear light source

$$I_d = I_Q \int_0^L \frac{\cos\theta}{r^2}\, \cos\alpha dl \cdot R_d$$

(c) Polygon light source

$$I_d = \frac{I_s}{2} \cdot \sum_{i=1}^{m} \gamma_i \cdot \cos\beta_i \cdot R_d$$

FIGURE 1. Calculation of illuminance due to various shapes of light sources.

over the illuminance due to all the point light sources of which it is composed (See Fig. 1 (b)).

c) Area and Polyhedron Light Sources

Extending the idea used for linear light sources, the illuminance due to an area light source can be obtained by integrating over the contour of the light source area as viewed from the calculation point (See Fig. 1 (c)).

In the same manner, illumination due to polyhedral sources can be calculated by integrating over the contour of each face of the polyhedral source, as viewed from the calculation point.

2.2 Shadowed Regions

To determine whether or not a convex polyhedron lies in the shadow of another convex polyhedron, we use the shadow volume formed by the polyhedra and the light source.

a) Point sources

A convex polyhedron and a point light source determine a single shadow volume, which forms an umbra on any face inside the volume.

b) Area Sources

Area sources form both umbrae and penumbrae, which can be calculated by extending the idea used for point light sources; the penumbra volume is determined by the

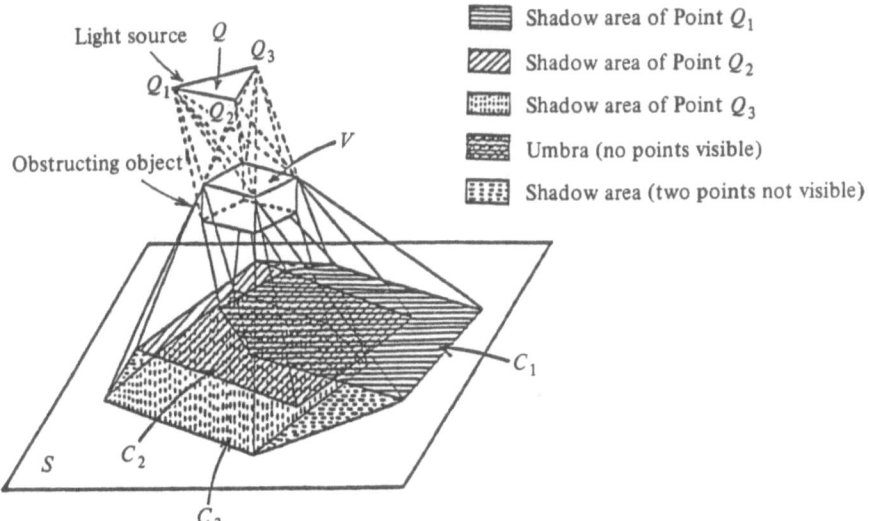

FIGURE 2. Umbra and penumbra regions for an area light source.

minimum convex volume containing all the shadow valumes, while the umbra volume is determined by the intersection of all the shadow volumes (See Fig. 2).

Shadows are formed on any face inside the penumbra or the umbra volume.

Caclulation of illuminance for points within the umbra volume is simple. If the point under consideration falls inside at least one umbra volume, then the direct illuminance is zero. On the other hand, inside penumbra volumes, the direct illumination is partially obstructed by at least one surface. Therefore, the illuminance at a point which falls inside at least one penumbra volume, but outside of all umbra volumes, must be calculated by obtaining the visible parts of the light source as viewed from the point (See Fig. 3).

2.3 Examples

Figure 4 (a) shows a room which is illuminated by a point light source, considering the distribution of luminous intensity. Figure 4 (b) is an example of a linear light source and a spotlight.

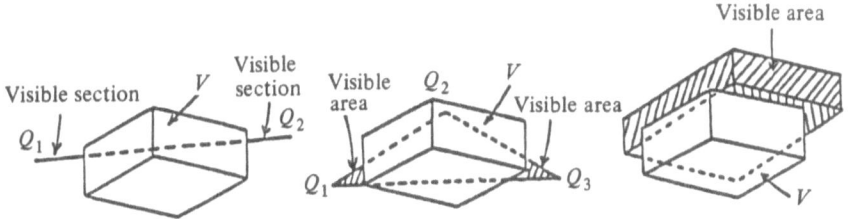

FIGURE 3. Visible parts of a light source viewed from the calculation point.

(a)   (b)

FIGURE 4. Examples. (Color art for this figure may be seen in the color insert.)

3 Mutual Inter-Reflection[3]

The effect of shadows on mutual inter-reflection is an important factor in rendering interior scenes. Most systems currently model indirect illumination by a uniform ambient light, ignoring the complex effects produced by mutual inter-reflection.

3.1 Fundamental Algorithm

a) Calculation of shadows on surfaces for each point is very inefficient. Considerable compute time can be saved by calculating the shadow areas (umbrae and penumbrae) before hidden surface removal Shadows are calculated in two steps as follows: First, umbra volumes and penumbra volumes formed by convex polyhedra interrupting illumination from a light source are obtained. Next, umbrae and penumbrae on each face are obtained as the intersection of the face with the umbra or penumbra shadow volumes

b) Mutual inter-reflection of light is calculated as follows: All object faces in the scene are subdivided into subfaces. Inter-reflection is calculated at the vertices of each subface before scan conversion. During scan conversion, the intensity of each point on a face is obtained by interpolation (See Fig. 5).

3.2 Examples

Figure 6 shows a computer room illuminated by two rectangular sources (ceiling lamps). In (a), only direct illumination has been calculated, while in (b), both direct and indirect illumination have been calculated. In both cases, shadows were also calculated. These pictures make clear that calculation of indirect illumination and of shadows is indispensable for generating realistic images.

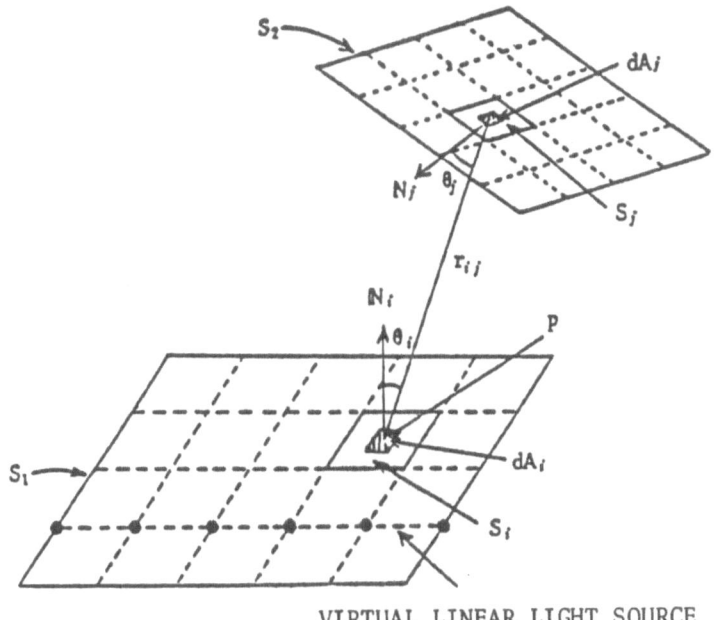

VIRTUAL LINEAR LIGHT SOURCE

$$I_i = I_{0i} + R_{di} \frac{I_j}{A_i} \int_{A_j} \int_{A_i} \frac{\cos\theta_i \cos\theta_j}{\pi r_{ij}^2} \, dA_i dA_j$$

FIGURE 5. Interreflection.

(a)   (b)

FIGURE 6. Examples. From *Computer Graphics*, Volume 19, Number 3, 1985. Reprinted with permission. (Color art for this figure may be seen in the color insert.)

162

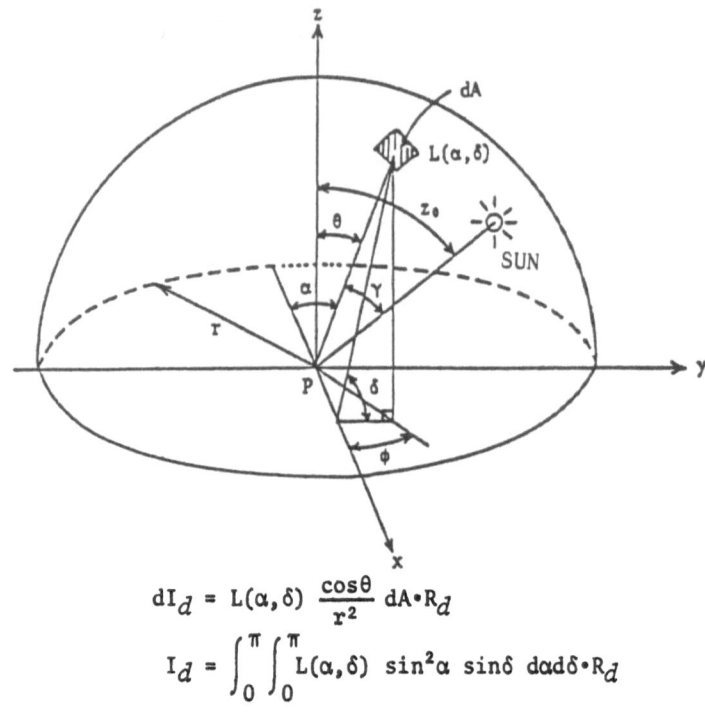

$$dI_d = L(\alpha,\delta)\ \frac{\cos\theta}{r^2}\ dA \cdot R_d$$

$$I_d = \int_0^\pi \int_0^\pi L(\alpha,\delta)\ \sin^2\alpha\ \sin\delta\ d\alpha d\delta \cdot R_d$$

FIGURE 7. Sky light effects.

4   Sky Light [4]

Natural lighting models to date have been limited to calculation of direct sunlight. Here we discuss calculation of natural lighting, considering both direct sunlight and sunlight scattered by clouds and other forms of water vapor in the atmosphere. This indirect natural light is termed skylight and is an important factor in rendering realistic images, particularly under overcast skies.

4.1   Fundamental Algorithm

a) The sky is supposed to be a hemisphere of very large radius (called the sky dome) which behaves as a source of diffuse light with non-uniform intensity (see Fig.7).

b) In order to account for the non-uniform intensity of the skylight, the sky dome is divided into bands.

c) The intensity within each band is assumed to be uniform in the transverse direction and nonuniform in the longitudinal direction. Thus, the total luminance due to each band can be accurately computed

4.2   Examples

Figure 8 shows some outdoor scenes including buildings and cars: (a) is shown with overcast skies, and (b) is shown with a clear sky with spotty clouds.

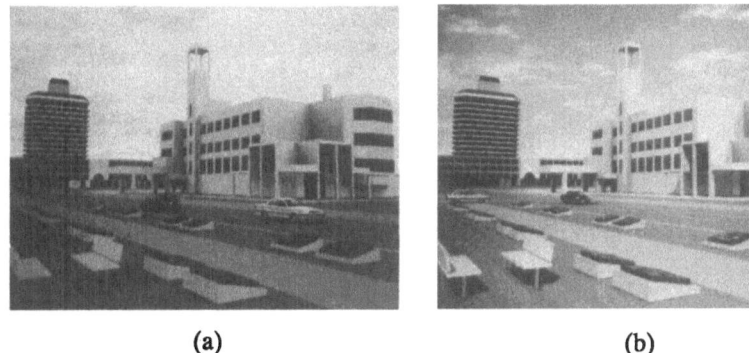

(a)   (b)

FIGURE 8. Examples. (Color art for this figure may be seen in the color insert.)

The calculation of these images includes mutual reflections from the ground and other objects in the scene. In other systems to date, shadows under overcast skies have been ignored. As seen in these images, the edges of shadows are not sharp, due to the presence of penumbrae. This is an imporatant effect for generating realistic images.

5 Atmospheric Scattering [5]

Consideration of scattering and absorption of light due to particles in the atmosphere is another important factor in generating realistic images.

5.1 Calculation of Light Scattering

Illuminated volumes in air glow due to light scattered from particles contained in the air. Furthermore, light is attenuated as it traverses a volume of air, due to absorption. These effects can be modeled as follows: For scattering, particles are considered to be an infinite number of point sources. For attenuation, the atmosphere is supposed to be a semitransparent object (see Fig. 9).

5.2 Examples

Figure 10 (a) shows a scene illuminated by four spotlights and a special effects light that combines a daisy shape and polka dots. The front left spotlight exhibits a soft edge, while the front right spotlight exhibits a sharp edge.

Picture (b) shows a building illuminated by searchlights, street lamps, and car headlights. Picture (c) is the same scene, with no fog in the air. Picture (d) shows a shaft of sunlight pouring in through a window.

light source

Q

$I(\theta,\phi)$

θ

r

s

P_v

viewpoint

illumination
axis

L

P

α

P_i

$$I = I_i e^{-\tau(L)} + \int_0^L I_p(s) e^{-\tau(s)} \sigma ds$$

FIGURE 9. Intensity of light reaching a viewpoint from a point P_i on an object. From *Computer Graphics*, Volume 21, Number 4, 1987. Reprinted with permission.

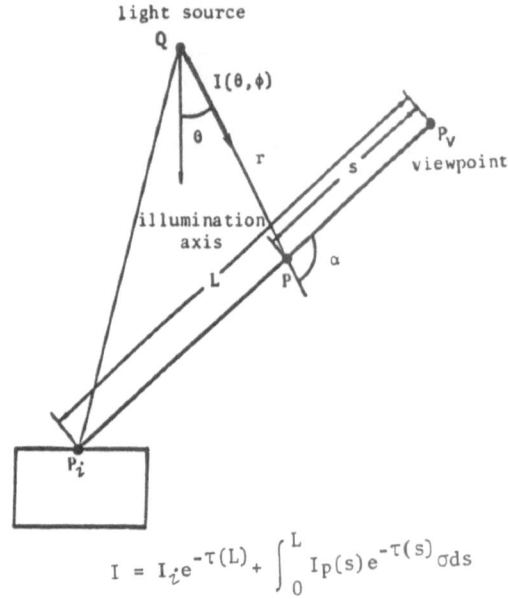

(a)

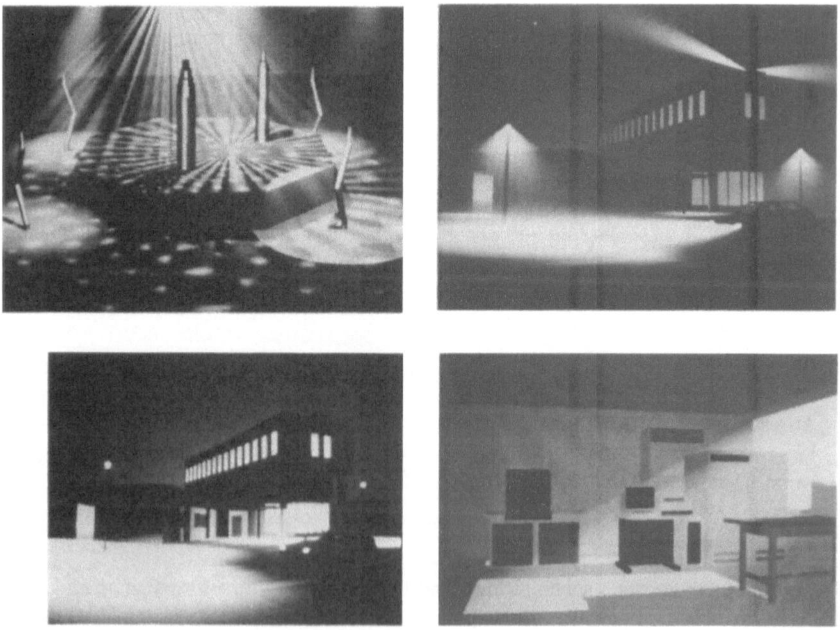

(c)

FIGURE 10. Examples. (a) and (d) are from *Computer Graphics*, Volume 21, Number 4, 1987. Reprinted with permission. (Color art for this figure may be seen in the color insert.)

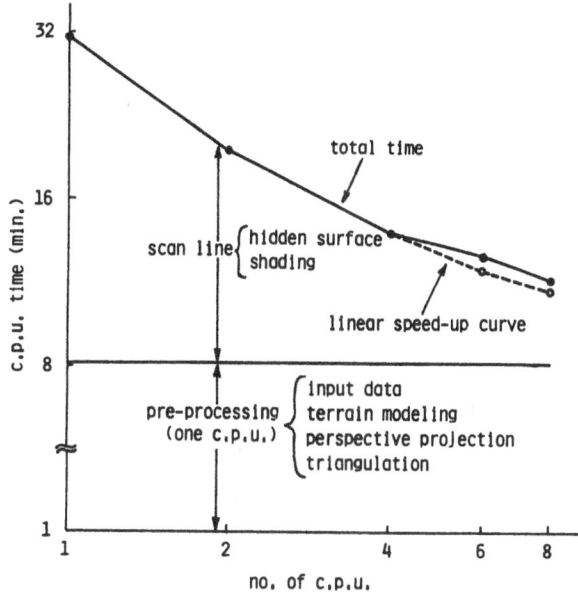

FIGURE 11. Efficiency of parallel processing applied to the scan line algorithm.

6 Computation Time

Scan line rendering is particularly amenable to parallel computation because the bulk of the processing time is absorbed in calculating shading, which can be done independently for each scan line.

The Symmetry, a multi-processor computer used in our laboratory, contains 12 microprocessors. In calculating a single image, we get nearly linear speedup when using multiple processors for the shading step only (See Fig. 11). For calculating animations, separate images are distributed to each processor; twelve images are simultaneously calculated.

References

[1] Nishita T. and Nakamae E.: "Half-Tone Representation of 3-D Objects Illuminated by Area Sources or Polyhedron Sources," IEEE COMPSAC, pp. 237-242 (1983)

[2] Nishita T., Okamura I. and Nakamae E.: "Shading Models for point and Linear Sources," ACM Transactions on Graphics, Vol. 4, No. 2, April, pp. 124-146 (1985)

[3] Nishita T. and Nakamae E.: "Continuous Tone Representation of Tree-Dimensional objects Taking Account of Shadows and Interreflection," Computer Graphics, Vol. 19, No. 3, pp. 23-30 (1985)

[4] Nishita T. and Nakamae E.: "Continuous Tone Representation of Tree-Dimensional Objects Illuminated by Sky Light," Computer Graphics, Vol. 20, No. 4, pp. 125-132 (1986)

[5] Nishita T., Miyawaki Y. and Nakamae E.: "A Shading Model for Atmospheric Scattering Considering Luminous Intensity Distribution of Light Sources," Computer Graphics, Vol. 21, No. 4, pp. 303-310 (1987)

Appendix

Appendix A

Co-processing Environments for Interactive Visualization

Craig Upson
Stellar Computer

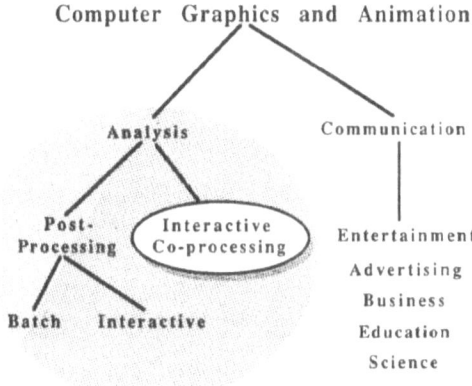

From the Institute for Supercomputing Research's second workshop on "Visualization in Supercomputing," August 22 to 25, 1988.

Computational Processes

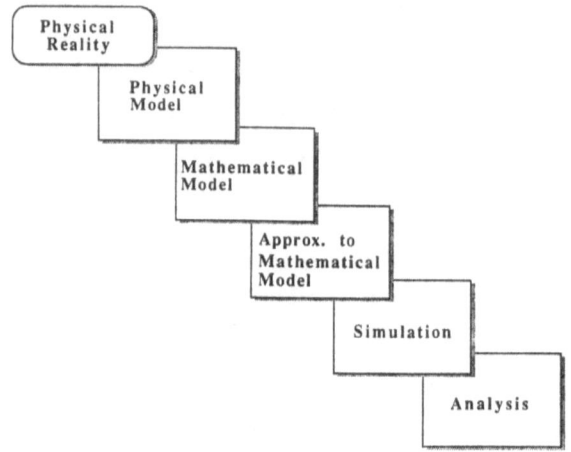

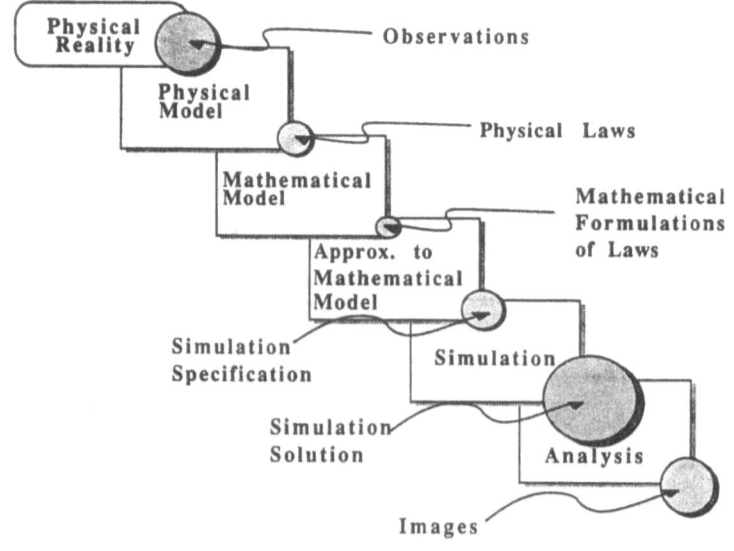

Computational Cycle

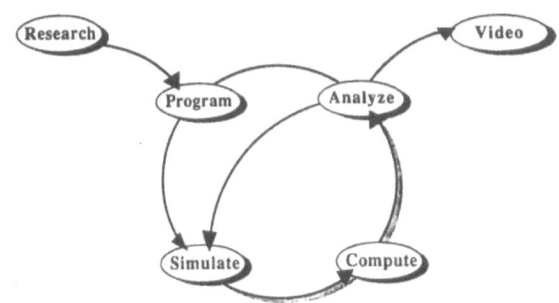

Analysis Cycle

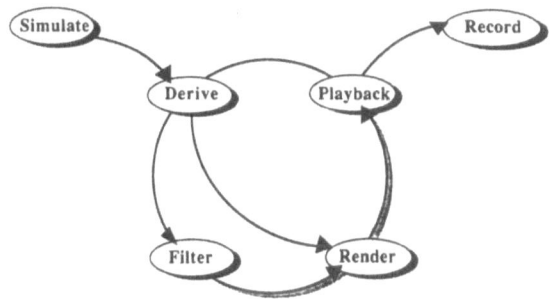

Analysis Cycle

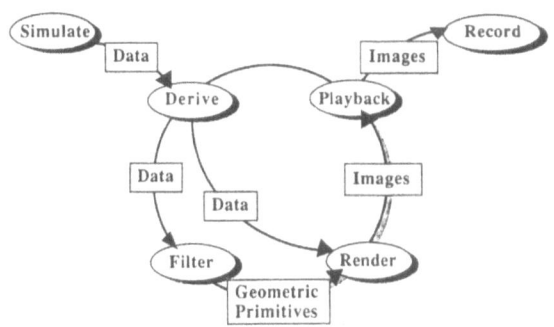

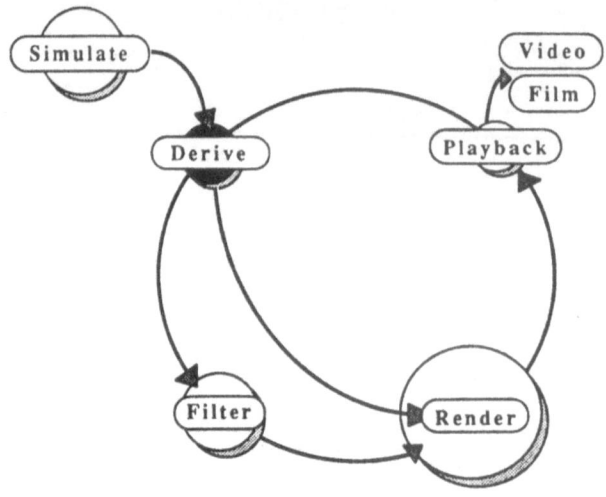

Data Manipulation and Filtering

- **Extraction:** Data maintains its type only
  selected attributes have changed

- **Derivation:** Data of a different type is derived
  from the original data

- **Geometric Filtering:** Deriving geometric
  primitives from (non-geometric) data

Data Extraction Processes:

- Extract   sub-regions
- Spatial or temporal sampling
- Spatial or temporal interpolation
- Boolean operations
- Copy about symmetry planes
- Extract N-M dimensional domain from N space
- Scale up/down
- Warp, Shear
- Image processing...smoothing, enhancement..
- Polygonal object subdivision
- etc ....

Data Derivational Processes:

- Compute statistical properties
- Compute gradient of a scalar field
- Derive streamfunction from velocity field
- Derive vorticity from velocity
- Advect particles from flow field
- Derive 3D texture map from scalar field
- etc ....

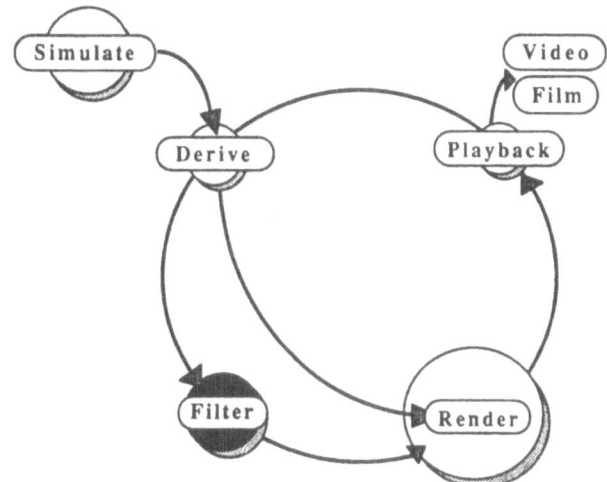

Geometric Filtering Processes:

- Derive Point Primitives
  - Scatter plots
  - Particles

- Derive Linear Primitives
  - Contour lines from 2D fields
  - Velocity vectors
  - Wireframes
  - Ribbon Molecular backbones

- Derive Surface Primitives
  - Point surfaces
  - Line surfaces
  - Polygonal surfaces
  - Nurb surfaces
  - Connolly molecule surfaces
  - Continuous color contour maps

- Derive Volume Primitives
  - Voxel
  - Cell
  - Cuberilles
  - Super quadrics

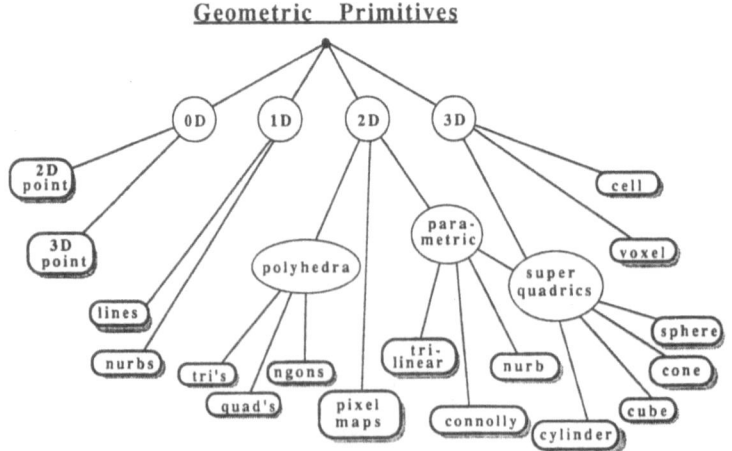

Geometric Primitives

Geometric Representations
of a Three Dimensional
Scalar Field

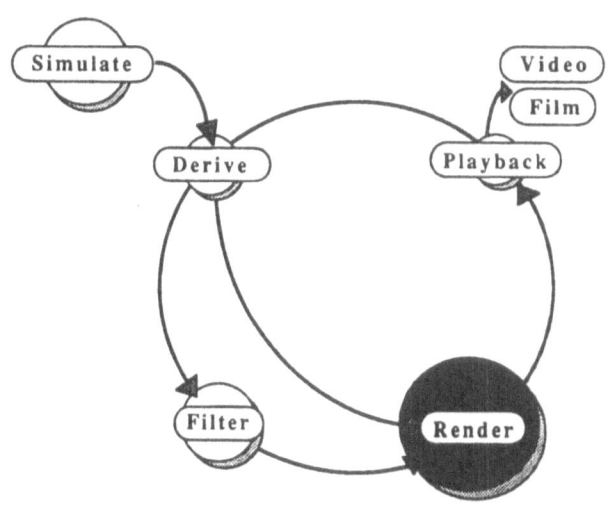

Geometric Attributes

- Object Hierarchy and Structure
  - Structure concatenation
  - Boolean selection operations
  - Spatial subdivision level
  - Antialiasing level

- Orientation
  - Position (translations, rotations)
  - Scale
  - Shape (shearing)

- Surface Responses to:
  - Ambient light
  - Diffuse light
  - Specular light
  - Surface roughness
  - Color
  - Reflectivity
  - Surface texturing

- Volumetric Properties
  - Transparency
  - Refractivity
  - Solid texturing

- Environmental Responses
  - Shadows
  - Global illumination (ray tracing)
  - Radiosity

Traditional Visualization Pipeline

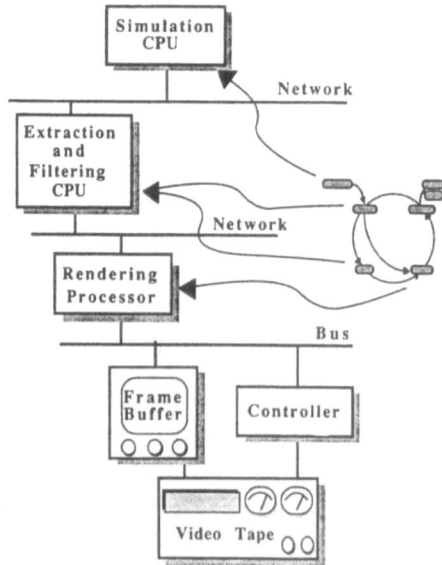

Traditional Visualization Pipeline

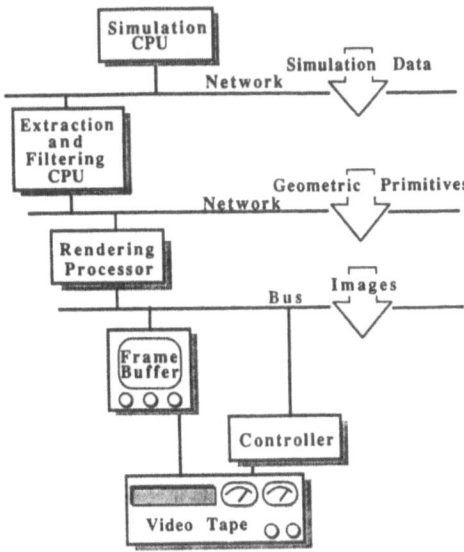

Traditional Visualization Pipeline

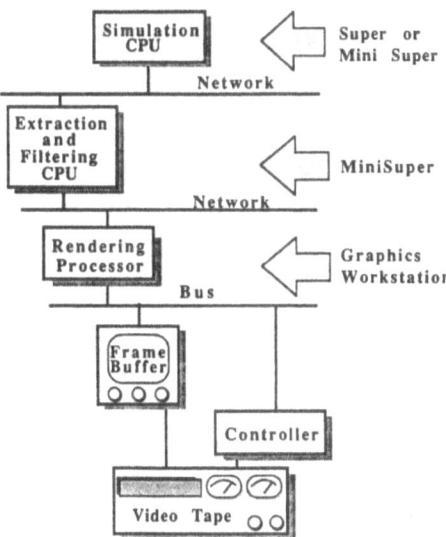

Traditional Visualization Bottleneck

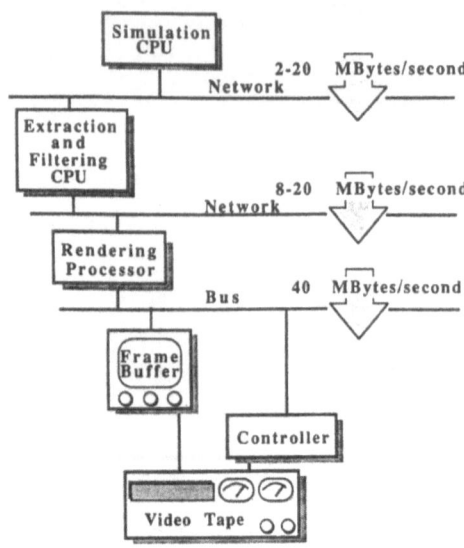

Interactive Visualization Pipeline

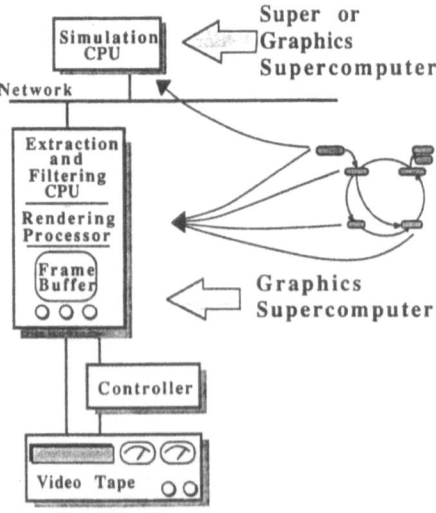

Visual Co-Processing

'....If the scientist cannot produce unambiguous and exhaustive formulations, and wishes instead to exercise his intuitive judgement as the calculation develops, he can arrange for that too. He can instruct the machine to present to him the relevant characteristics of the situation, continuously or in discrete succession, as the calculation progresses, by oscilloscopic graphing. He can then intervene whenever he sees fit.'

John von Neumann - 1946

Requirements for Interactive Visualization:

Hardware Performance

Visualization Environments

A Typical 3D Simulation
Appropiate for Graphics Supercomputing

- Medium Sized 3D CFD Simulation:
Tornadic Storm Evolution

    - Finite Difference
    - 21,600 Nodes or Cells   (25x27x32)
    - 8 Variables per Node
    - 700 - 1000 Timesteps
    - One Timestep every 3 Seconds

- Visualization Goals:

    - Co-Processing *not* Post Processing
    - Must Keep up with Simulation
    - Must be Interactive
        - ~ 5-10 Frames per Second
    - 3D Visual Representation

Hardware Performance Issues

MFLOPs

Bandwidth

Memory

Rendering

Visualization Styles:

Velocity   Vectors

Passive   Particle   Tracers

Contour   Surfaces

Hardware Performance Issues:

MFLOPS

• **Simulation:**

  • 4 MFLOPs per Second
    (3 times faster than real time)

• **Derivation:**

  • 1 MFLOP per Second
    (for particle advection)

• **Geometric   Filtering:**

  • 2-4 MFLOPs per Geometric Object
    (for polygonal surfaces)

Hardware Performance Issues:

MFLOPS

• **Geometric Transformations:**

- **0.7 MFLOP per Second**
  (for velocity vectors)

- **0.2 MFLOP per Second**
  (for particles)

- **2-6 MFLOPs per Second**
  (for polygonal surfaces)

    Total: 10-15 MFLOPs per Second
    Sustained

Hardware Performance Issues

BANDWIDTH

• **Simulation Data:**

- **2 - 4 MBytes per Second**

• **Geometric Objects:**

- **2 - 4 MBytes per Second**

• **Images:**

- **20- 40 MBytes per Second**

    Total: 24 - 48 MBytes per Second
    Sustained

• **Peak Ratings must be between**
  240 - 480 MB/Second

Hardware Performance Issues:
MEMORY

- **Simulation Data:**

    - **0.7 MBytes per Timestep**
    - **Need Several Timesteps (3)**

- Geometric Objects:

    - **0.2 - 0.4 MBytes per Object**
    - **Need Several Objects (3)**

- Images:

    - **4 MBytes per Frame**
    - **Need Access to Several Frames
      to Compare Images (3)**

    ### Total: 16 - 32 MBytes of Memory

Hardware Performance Issues:
RENDERING

- Vectors:
    - **5 - 10 Frames per Second**
    - **400 - 800 Vectors per Frame**
    - **500 Operations per Vector**

- Spheres:
    - **5 - 10 Frames per Second**
    - **500 - 1000 Spheres per Frame**
    - **5000 Operations per Polygon**

- Polygonal Surfaces:
    - **5 - 10 Frames per Second**
    - **2000 to 5000 Polygons per Frame**
    - **1200 Operations per Polygon**

Total: 20 - 115 MIPs per Second for Pixel Calculations

Hardware Performance Issues:
RENDERING

- **Rendering Flexibility**

    - **Rendering Processors need to be Programmable**

        **We don't Currently know what Geometric
        Primitives we will need in the Future**

Visualization Environment Issues

- **Application Environments:**

    - **Transparently Perform Tasks not Directly Related to Solving the User's Problem:**

        - Hides Low Level Complexity

    - **Provide a Consistent View of the System to the User**

        - Consistent User Interface

    - **Encourage the Re-Cycling of Code Modules**

        - Build on what has been done before

    - **Provide a Data Repository for Comparing Prior Computations**

    - **Are an Integrated set of Tools**

Requirements for Visualization Environments

- **Ease of Use**

    - Productivity is as Important as Performance

- **Computational Completeness**

    - Graphics is only Part of the Problem

- **Extensible and Customizable**

    - No System can Solve all Problems

- **Encourages Rapid Prototyping**

    - Need to Rapidly Iterate on Designs

- **Builds Upon Standards for Portability**

    - Standard Graphics Libraries
    - Standard Windowing Systems

Appendix B

Supercomputing Environments for the 1990s

Larry Smarr
University of Illinois at Urbana Champaign
Director, National Center for Supercomputing Applications

NCSA MISSION

NCSA'S MISSION IS TO USE THE TOOLS AND
TECHNIQUES OF LEADING-EDGE
COMPUTATIONAL SCIENCE TO:

- CONTRIBUTE TO NEW ADVANCES IN
  SCIENCE AND ENGINEERING

- STRENGTHEN THE INDUSTRIAL
  COMPETITIVENESS OF AMERICAN
  INDUSTRY; AND

- CREATE NEW ADVANCES IN
  COMPUTATIONAL SCIENCE ITSELF.

From the Institute for Supercomputing Research's second workshop on
"Visualization in Supercomputing," August 22 to 25, 1988.

184

NCSA UNIX Pathway

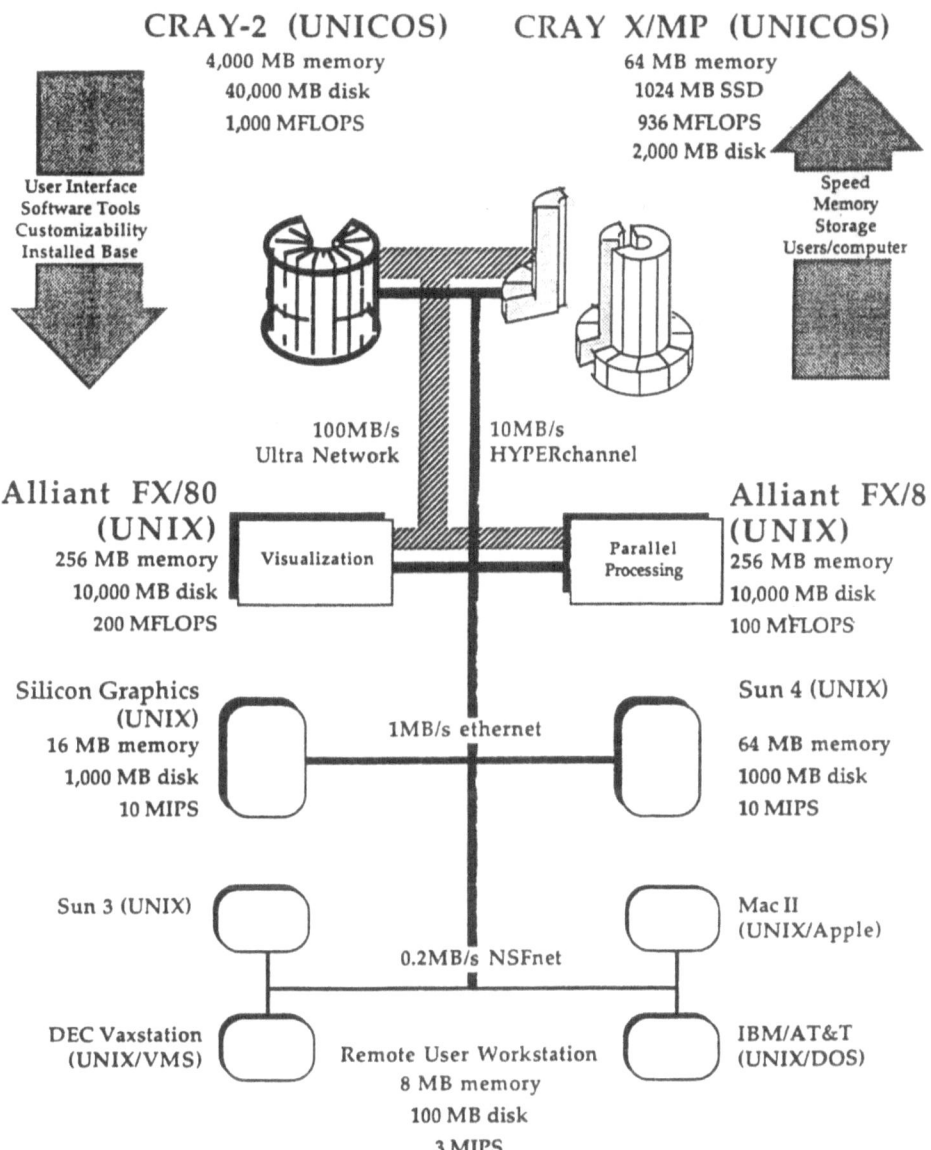

CRAY-2 (UNICOS)
4,000 MB memory
40,000 MB disk
1,000 MFLOPS

CRAY X/MP (UNICOS)
64 MB memory
1024 MB SSD
936 MFLOPS
2,000 MB disk

User Interface
Software Tools
Customizability
Installed Base

Speed
Memory
Storage
Users/computer

100MB/s
Ultra Network

10MB/s
HYPERchannel

Alliant FX/80 (UNIX)
256 MB memory
10,000 MB disk
200 MFLOPS

Visualization

Parallel Processing

Alliant FX/8 (UNIX)
256 MB memory
10,000 MB disk
100 MFLOPS

Silicon Graphics (UNIX)
16 MB memory
1,000 MB disk
10 MIPS

1MB/s ethernet

Sun 4 (UNIX)
64 MB memory
1000 MB disk
10 MIPS

Sun 3 (UNIX)

Mac II
(UNIX/Apple)

0.2MB/s NSFnet

DEC Vaxstation
(UNIX/VMS)

Remote User Workstation
8 MB memory
100 MB disk
3 MIPS

IBM/AT&T
(UNIX/DOS)

PROPOSED
COMPREHENSIVE SUPERCOMPUTING ENVIRONMENT

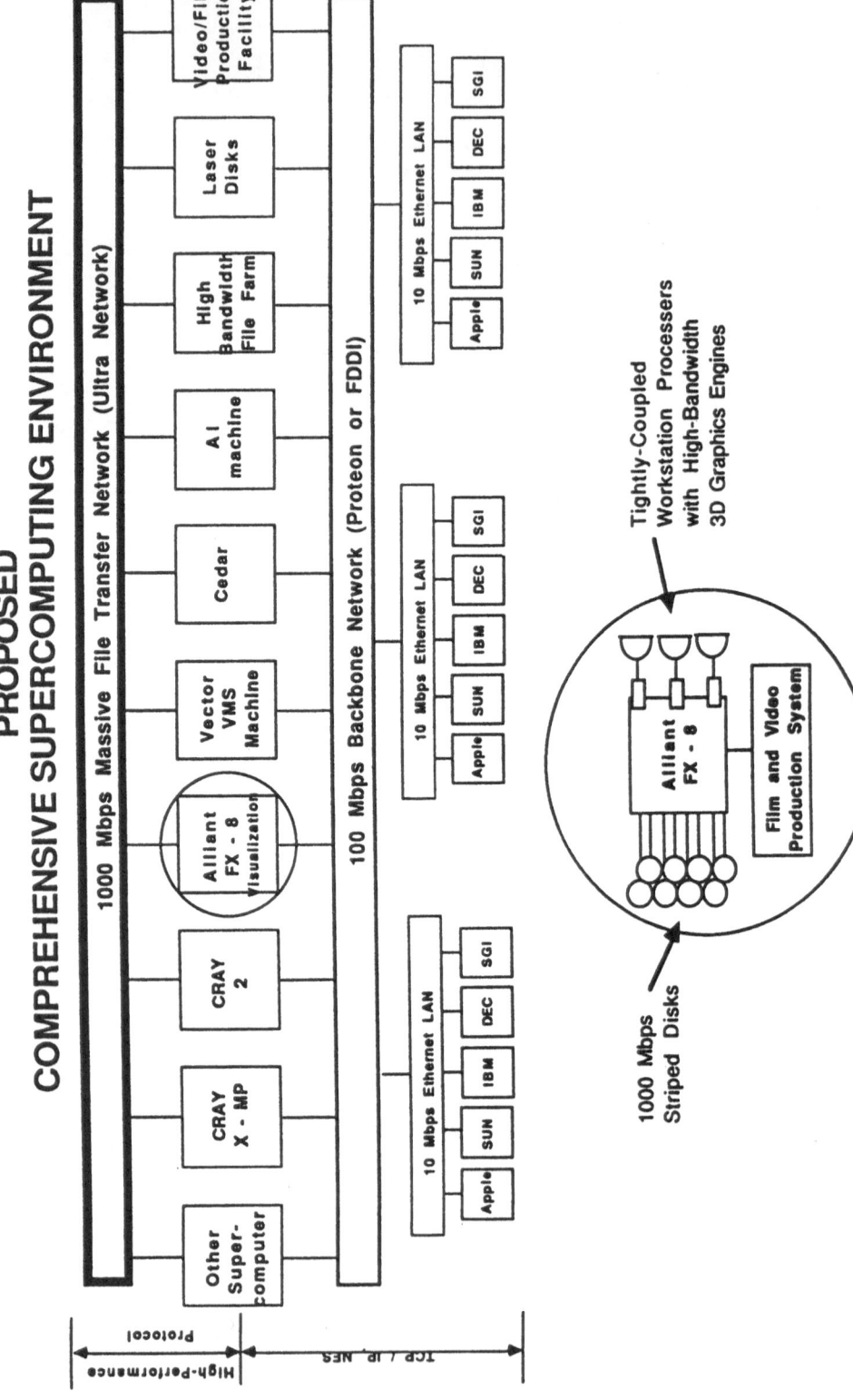

Planned FY89 Supercomputer Facility and Network

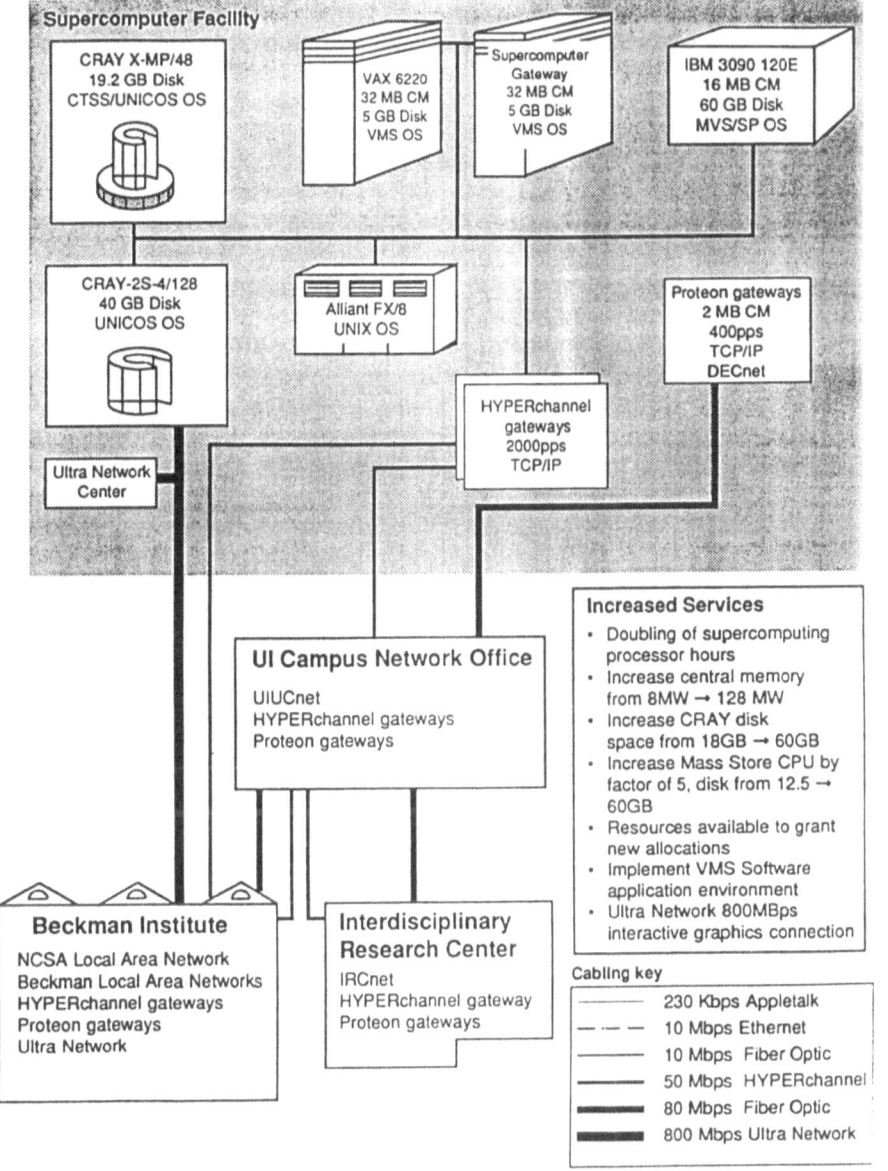

188

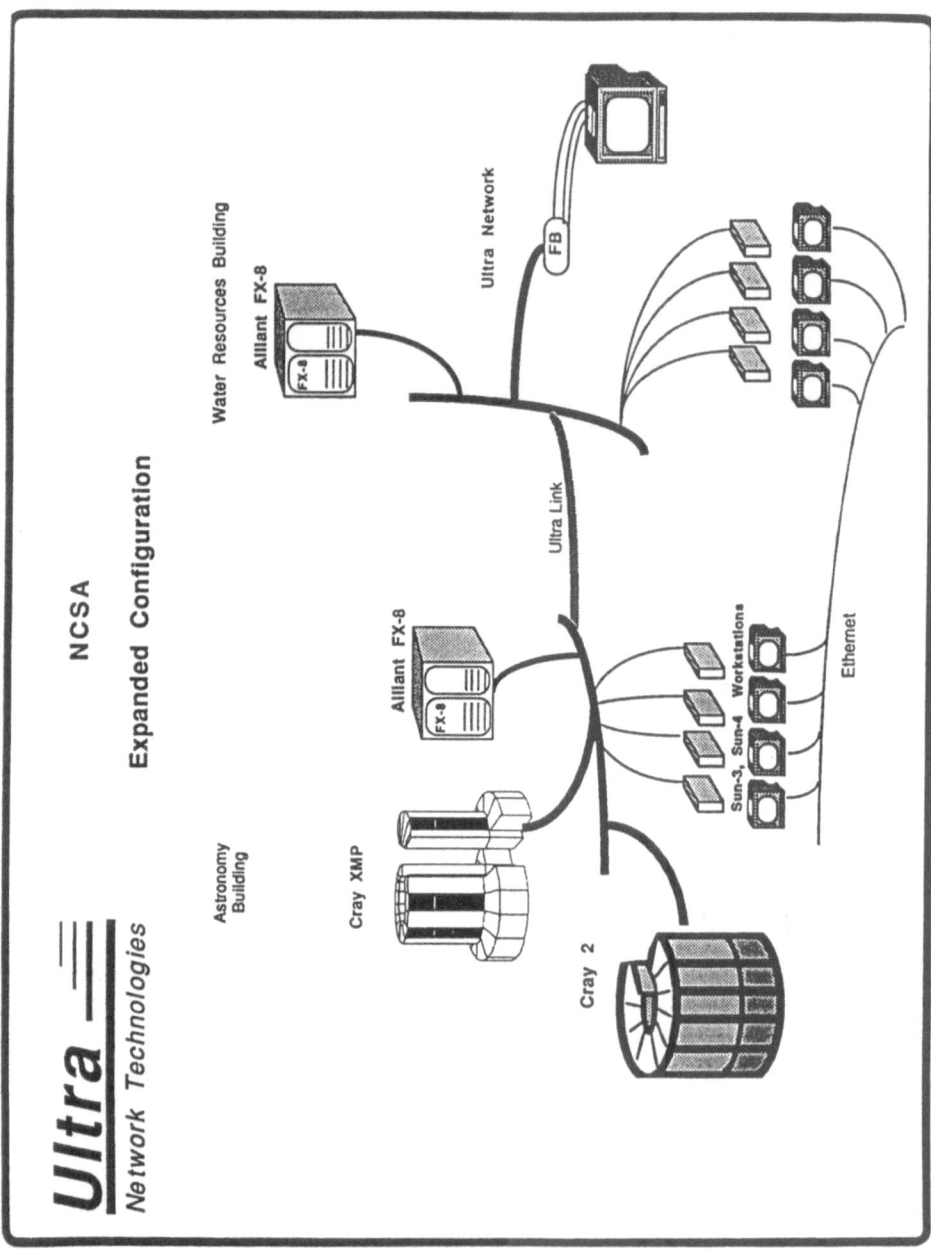

Planned FY89 Interdisciplinary Research Center IRCnet
Providing Network Services for NCSA Staff and Visitors

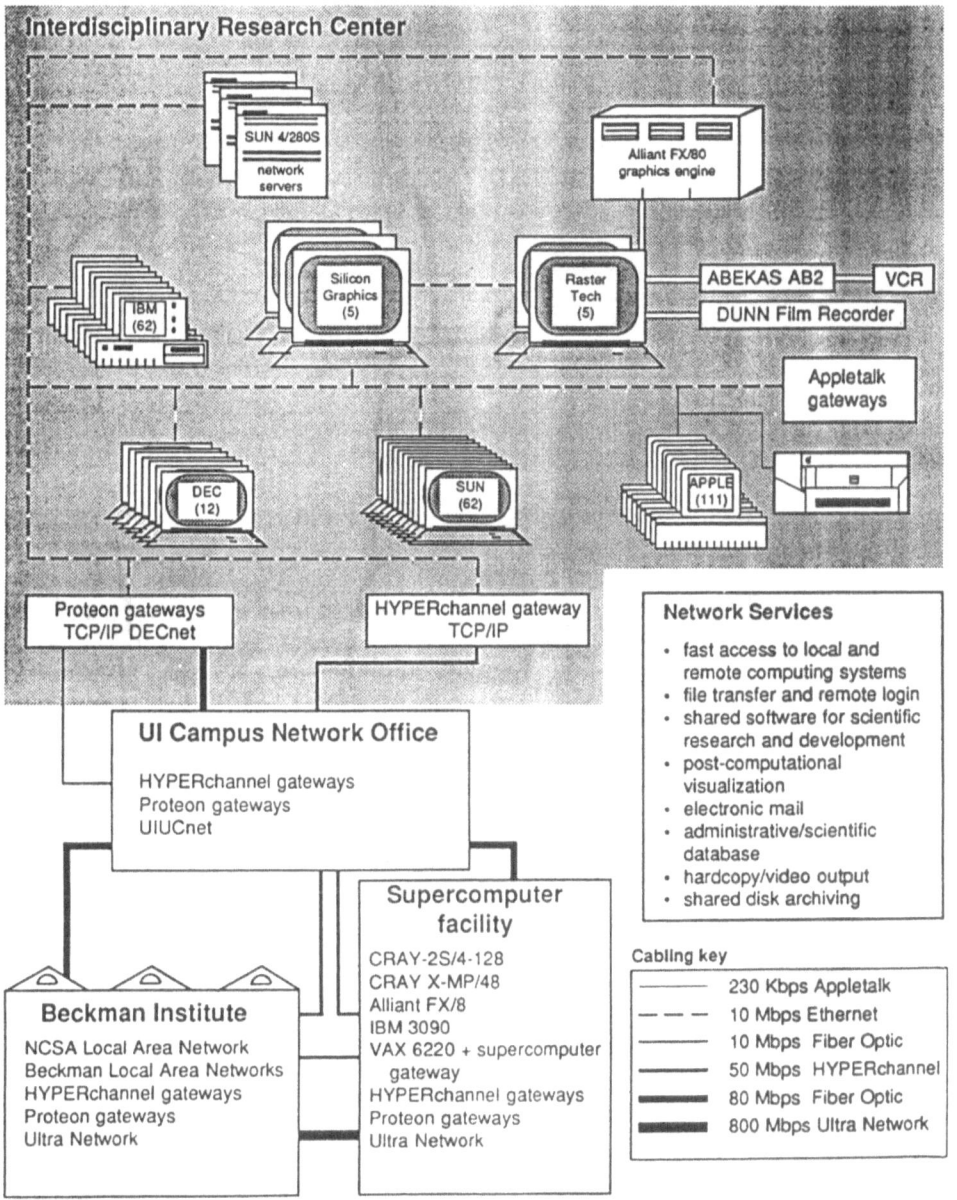

Interdisciplinary Research Center

SUN 4/280S
network
servers

Alliant FX/80
graphics engine

IBM
(62)

Silicon
Graphics
(5)

Raster
Tech
(5)

ABEKAS AB2 — VCR

DUNN Film Recorder

Appletalk
gateways

DEC
(12)

SUN
(62)

APPLE
(111)

Proteon gateways
TCP/IP DECnet

HYPERchannel gateway
TCP/IP

Network Services

- fast access to local and remote computing systems
- file transfer and remote login
- shared software for scientific research and development
- post-computational visualization
- electronic mail
- administrative/scientific database
- hardcopy/video output
- shared disk archiving

UI Campus Network Office

HYPERchannel gateways
Proteon gateways
UIUCnet

Supercomputer facility

CRAY-2S/4-128
CRAY X-MP/48
Alliant FX/8
IBM 3090
VAX 6220 + supercomputer gateway
HYPERchannel gateways
Proteon gateways
Ultra Network

Beckman Institute

NCSA Local Area Network
Beckman Local Area Networks
HYPERchannel gateways
Proteon gateways
Ultra Network

Cabling key

————	230 Kbps Appletalk
– – –	10 Mbps Ethernet
————	10 Mbps Fiber Optic
————	50 Mbps HYPERchannel
▬▬▬	80 Mbps Fiber Optic
▬▬▬	800 Mbps Ultra Network

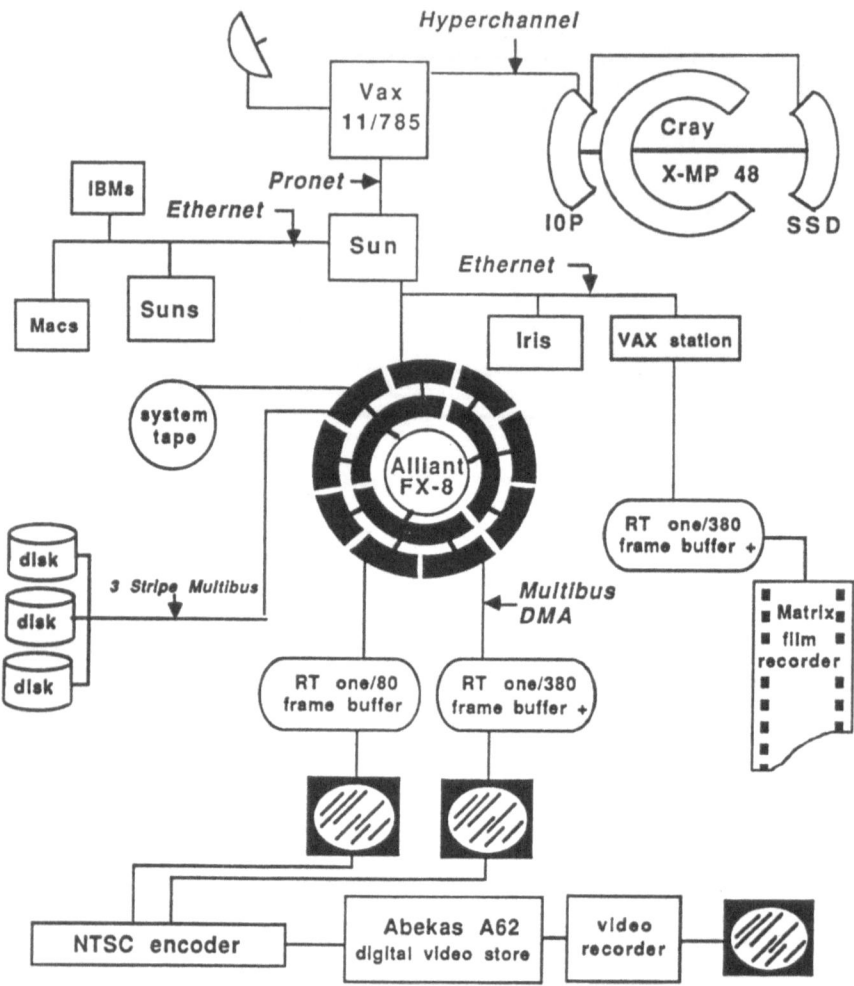

NCSA High Speed Visualization Facility, Phase 1

Scientific Media Services

RGB Environment

RGB Monitor

RGB Monitor

Digitizing camera
copy stand

RT One / 380
1280 x 1024
frame buffer

RT One / 80
1280 x 1024
640 x 480
frame buffer

Dunn 638
Analog Film
Recorder
1280 x 1024

Dunn 635
Analog Film
Recorder
640 x 480

DMA

DMA

**NTSC composite
Edit and dubbing
Environment**

Faroudja
Encoder

Lyon Lamb
ENC DI Encoder

Folsom RGB/NTSC
Scan Converter

Abekas A62
Digital Disk Recorder

SONY BUU850 UTR
Broadcast 3/4"

monitors

UHS   Beta   Sony 3/4"

Computer Environment

Cray
XMP48

Alliant FX / 8

IBM PC AT
AT&T Targa 16
512 x 512
frame buffer
Lumena Paint

Silicon Graphics
2400

high res RGB
workstations
SGI, Sun, Mac II, Ras Tech, etc.

Network

Key

DATA

RGB - 1280 x 1024

RGB - 640 x 480

RGB - 512 x 512

RGB - 512 x 512

NTSC Composite

National Center for Supercomputing Applications
Cumulative Usage By CPU Hours
January 86 - June 88

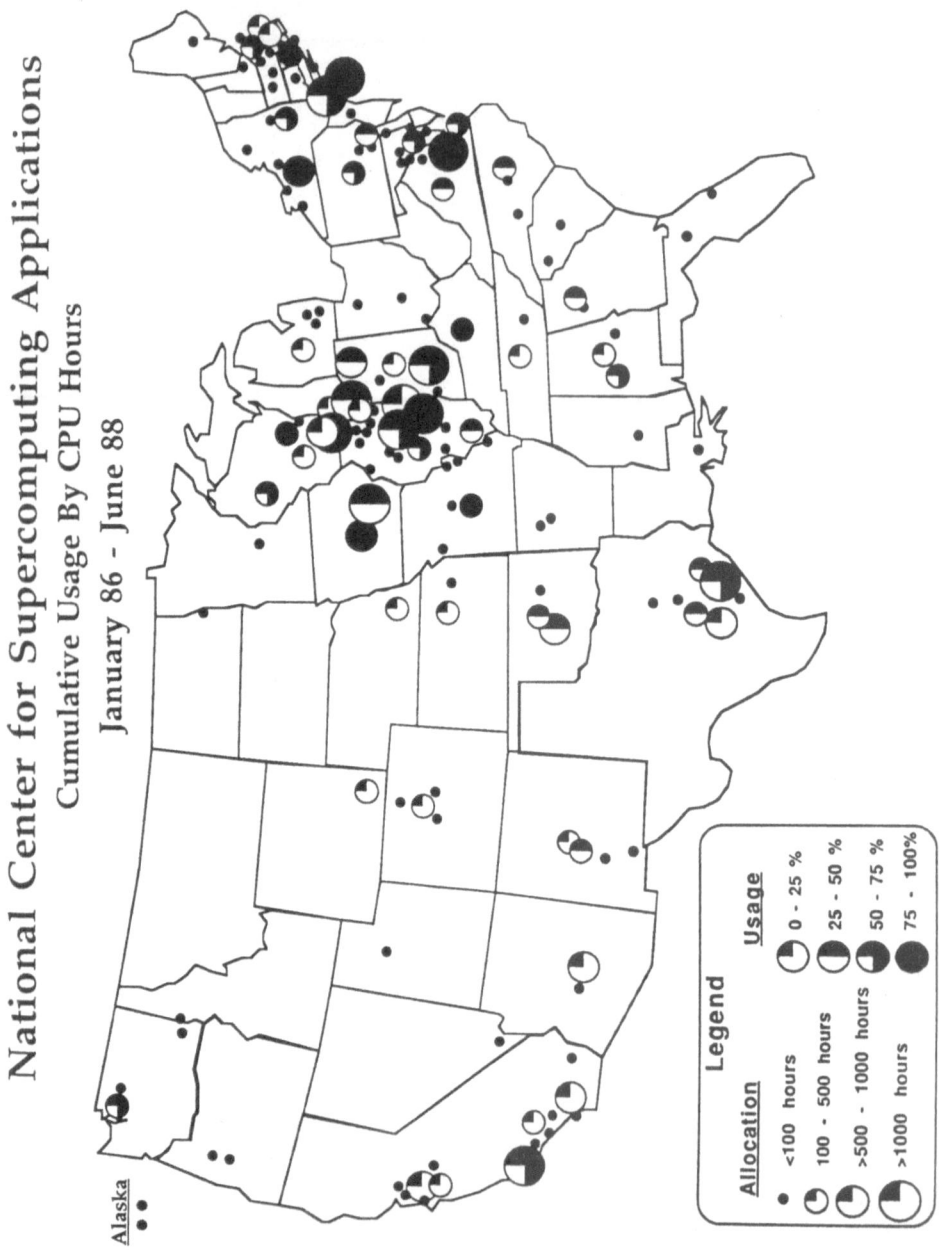

Alaska

Legend

Allocation

• <100 hours
◔ 100 - 500 hours
◔ >500 - 1000 hours
◔ >1000 hours

Usage

◔ 0 - 25 %
◑ 25 - 50 %
◕ 50 - 75 %
● 75 - 100%

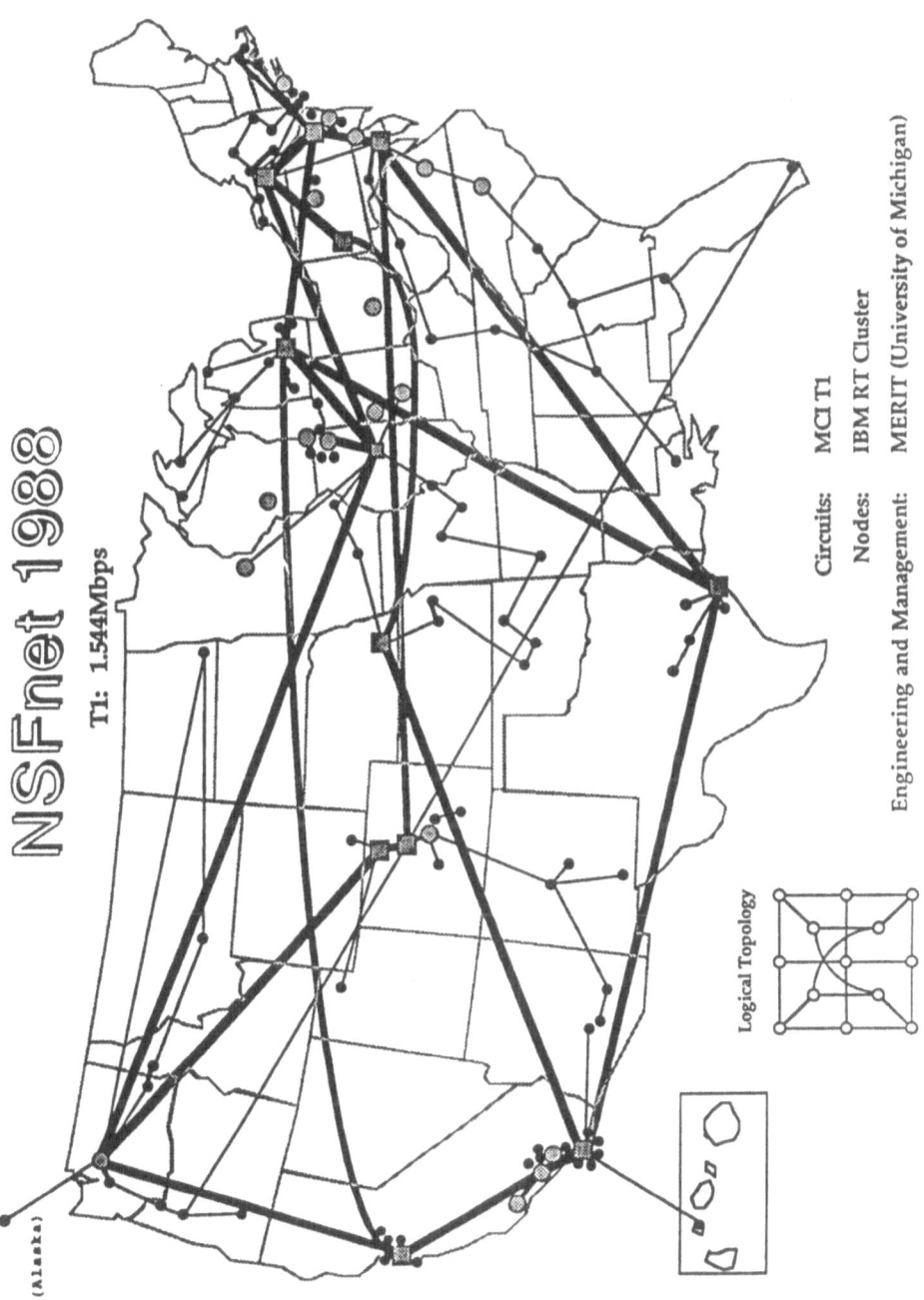

NSFnet 1988

T1: 1.544Mbps

Logical Topology

(Alaska)

Circuits: MCI T1

Nodes: IBM RT Cluster

Engineering and Management: MERIT (University of Michigan)

NCSA Support Network
Projected Growth
(Log scale)

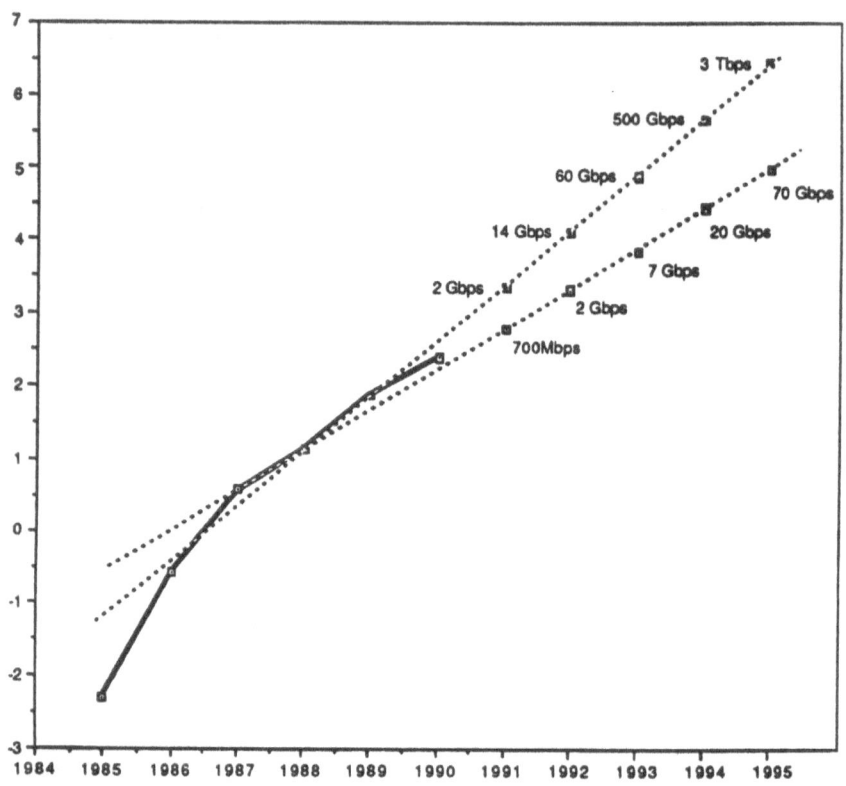

Remote File Transfer

Estimated Bandwidth Needed for
Single Users*

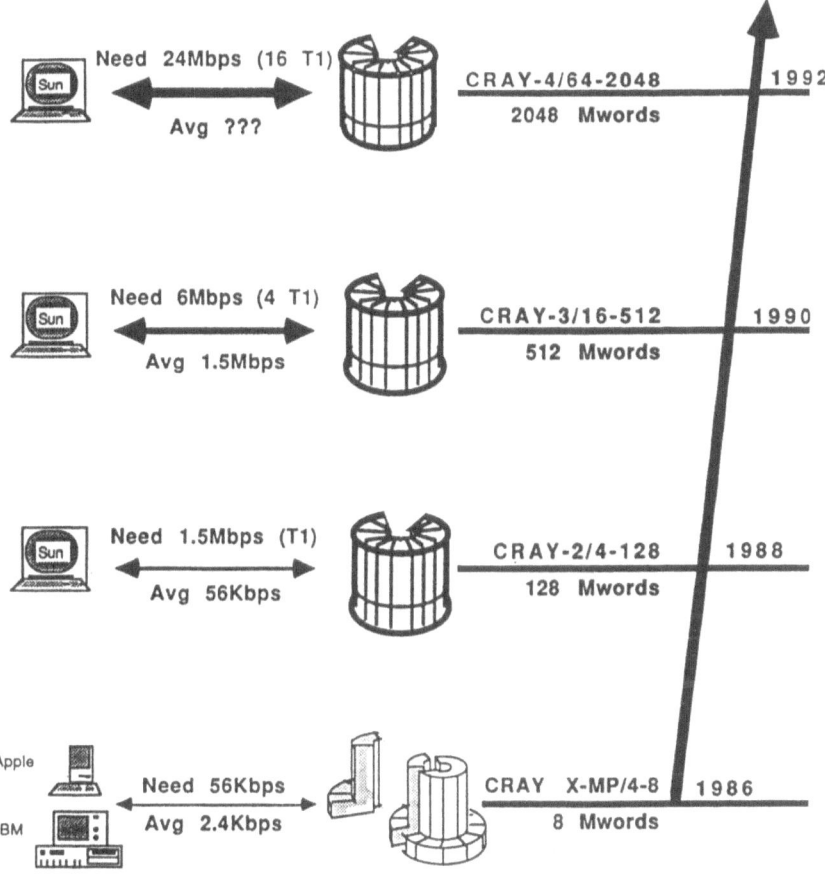

* Individual visualization needs may be higher.
  Based on avg file size 1/10 memory size

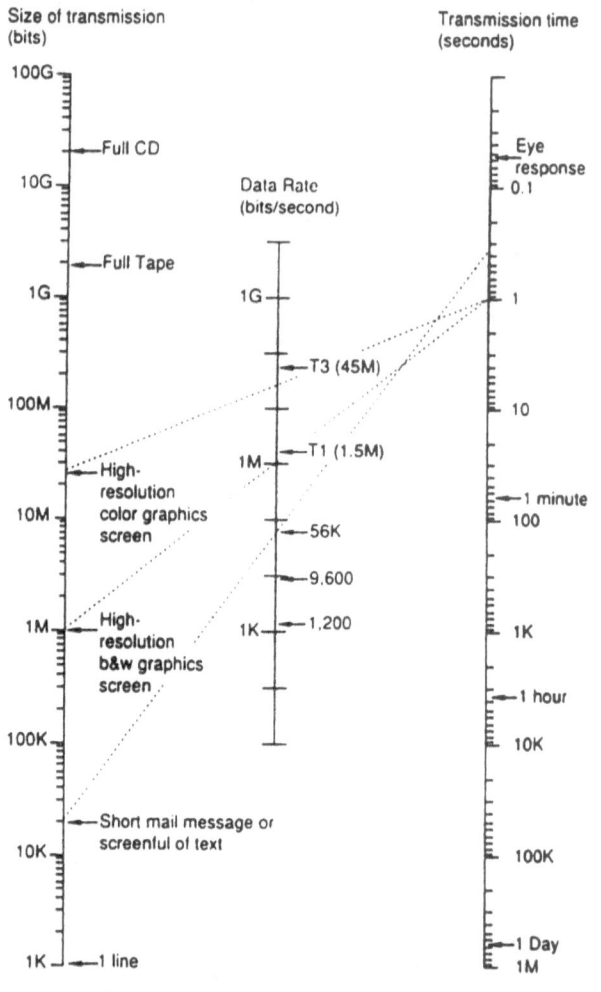

A nomogram deriving transmission time from transmission size and data rate.

NCSA's Distributed Computing Environment

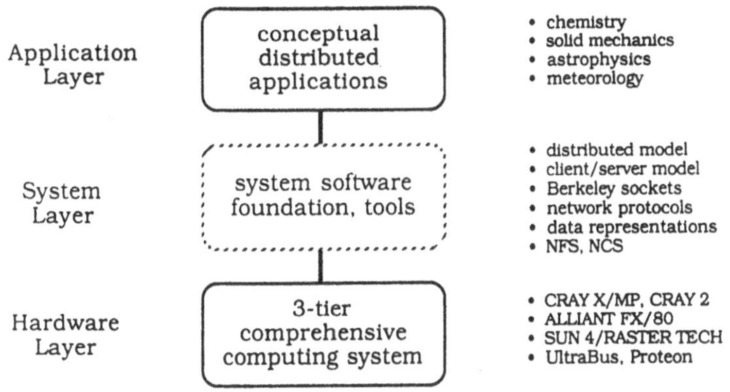

Application Layer	conceptual distributed applications	• chemistry • solid mechanics • astrophysics • meteorology
System Layer	system software foundation, tools	• distributed model • client/server model • Berkeley sockets • network protocols • data representations • NFS, NCS
Hardware Layer	3-tier comprehensive computing system	• CRAY X/MP, CRAY 2 • ALLIANT FX/80 • SUN 4/RASTER TECH • UltraBus, Proteon

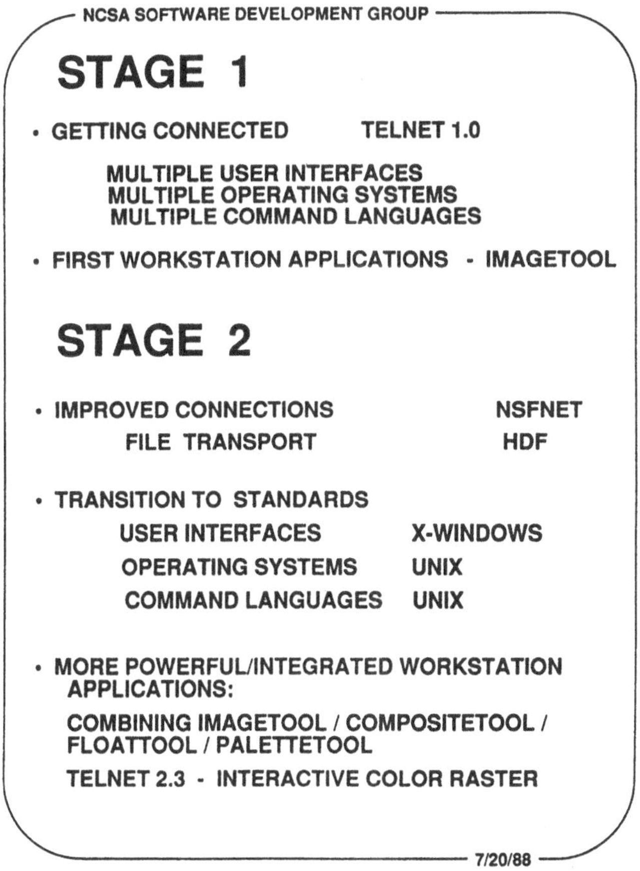

NCSA SOFTWARE DEVELOPMENT GROUP

STAGE 1

• **GETTING CONNECTED**   **TELNET 1.0**

   MULTIPLE USER INTERFACES
   MULTIPLE OPERATING SYSTEMS
   MULTIPLE COMMAND LANGUAGES

• **FIRST WORKSTATION APPLICATIONS - IMAGETOOL**

STAGE 2

• **IMPROVED CONNECTIONS**   **NSFNET**
   FILE TRANSPORT   **HDF**

• **TRANSITION TO STANDARDS**
   USER INTERFACES   **X-WINDOWS**
   OPERATING SYSTEMS   **UNIX**
   COMMAND LANGUAGES   **UNIX**

• **MORE POWERFUL/INTEGRATED WORKSTATION APPLICATIONS:**
   COMBINING IMAGETOOL / COMPOSITETOOL / FLOATTOOL / PALETTETOOL
   TELNET 2.3 - INTERACTIVE COLOR RASTER

7/20/88

198

What Scientists need to do their work

- **Connectivity** - access to supercomputing and other resources across available networks

  NCSA Telnet

- **Flexible file capability** - compatibility across machines and systems for data and results

  NCSA Hierarchical Data Format

- **Image analysis** - view and manipulation of color raster images on local workstation

  NCSA ImageTool

- **Presentation tools** - communication of results of research to colleagues in clear format

  NCSA CompositeTool

STAGE 3

- **ENHANCING COMMUNICATIONS / CONNECTIVITY**

  WIDE AREA - T1 NSFNET / SWITCHED 56KB

  LOCAL - HSC

  HDF - STORAGE / TRANSPORT / COMPATIBILITY

- **DEVELOPMENT AROUND STANDARDS**

  UNIX / X-WINDOWS OPTIMIZED FOR HARDWARE / HDF

- **BEGINNING OF TRUE DISTRIBUTED PROCESSING TOOLS**

  NETWORK COMPUTING FORUM - PARTICIPATION IN SETTING INDUSTRY-WIDE STANDARDS

Scope of the RIVERS Project

A. Hardware Research

* Three-tier computing environment

* Real-time, interactive visualization

* High-performance (>Gigabit/sec) networks

* Advanced Mass-Storage Devices

* Advanced video production facilities

B. Software Research

* Interactive, real-time visualization and animation software

* User interfaces and UIMS

* Run-time monitoring and interactive steering of simulations

* High-performance distributed scientific computing systems

    - Memory-to-memory oriented communications

    - Shared memory, multi-tasking between hosts

    - Integrated, multi-vendor software systems (e.g. binary floating point formats)

    - Object-oriented software development environment

    - "Bread boarding" applications - configuration-independent software systems

* Integration with data base; parallel processing and symbolic computing systems

Computational Scientist's Workbench

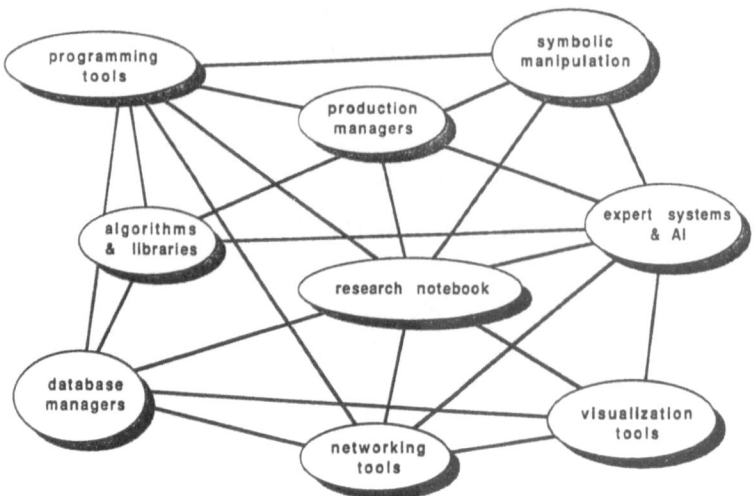

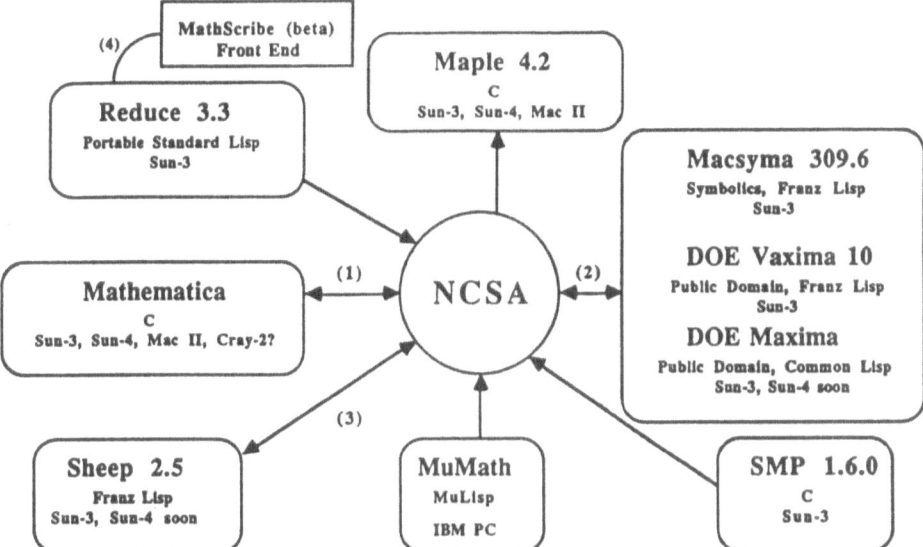

Symbolic Manipulation at NCSA

(1) Mathematica, Sun-4 port site, Tensor program (with L.Parker).
(2) Advisors to Symbolics, Port of Common Lisp Maxima to SUN-3 and SUN-4.
(3) To assist with SUN port of Sheep and Stensor.
(4) Beta test site for MathScribe friendly front end to Maple and Reduce from Tektronix

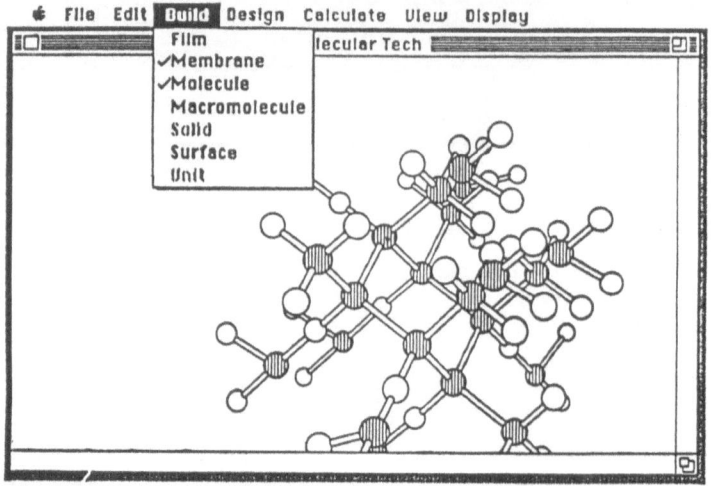

Build menu:
- Film
- ✓Membrane
- ✓Molecule
- Macromolecule
- Solid
- Surface
- Unit

Menu bar: ■ File Edit **Build** Design Calculate View Display

Window title: ...lecular Tech

Design
- Assemble
- Guess
- Model

Calculate
- Molecular Mechanics
- Electronic Structure
- Functions
- Properties
- Dynamics
- Monte Carlo

View
- Animate
- Dynamics
- Molecule
- Properties